Environment and Society

Debra J. Davidson · Magnus Boström ·
Pedro Roberto Jacobi
Editors

Environment and Society

Concepts for a Just Transformation

palgrave
macmillan

Editors
Debra J. Davidson
University of Alberta
Edmonton, AB, Canada

Magnus Boström
Örebro University
Örebro, Sweden

Pedro Roberto Jacobi
Universidade de São Paulo
Sao Paulo, Brazil

ISBN 978-3-032-27058-0 ISBN 978-3-032-27059-7 (eBook)
https://doi.org/10.1007/978-3-032-27059-7

Cover illustration: northlightimages

This Palgrave Macmillan imprint is published by the registered company Springer Nature Switzerland AG
The registered company address is: Gewerbestrasse 11, 6330 Cham, Switzerland

In memory of C.S.A. Kris van Koppen, who showed us that knowledge grows best when shared with generosity, curiosity, and joy.

Contents

Notes on Contributors

Mufti Nadimul Quamar Ahmed is a Ph.D. candidate in environmental sociology at Utah State University. His academic research interests include climate change perception, fertility decisions and migration as forms of adaptation, and environmental justice, with a focus on the lived experiences of marginalized populations, including environmentally vulnerable populations in Bangladesh and immigrant communities in the United States.

Ana Terra Amorim-Maia Ph.D., is an interdisciplinary socio-environmental scientist working on urban climate governance, adaptation, and justice. Her research focuses on how cities design, implement, and evaluate equitable climate adaptation, with particular expertise in intersectional climate justice. She collaborates closely with city governments and international organizations to bridge research, policy, and practice.

Isabelle Anguelovski is an ICREA Research Professor at ICTA-UAB and current Director of the Barcelona Lab for Urban Environmental Justice and Sustainability (BCNUEJ). Her work focuses on urban environmental justice, green and climate gentrification, just housing, and nature-based solutions, with strong community and policy engagement.

José Barrena is an Associate Researcher at the Transdisciplinary Center for Environmental Studies and Sustainable Human Development (CEAM) at the Austral University of Chile. His work examines socio-ecological transformations linked to aquaculture expansion, nature conservation, nature-based tourism, and Indigenous territorial claims, with a particular focus on Chilean

Patagonia. He has published on marine frontiers, ocean governance, and conflicts over coastal and maritime spaces in the context of global environmental change.

Magnus Boström is Professor of Sociology in the School of Humanities, Education, and Social Sciences at Örebro University, Sweden, with a research profile in environmental sociology, sociology of consumption, and political sociology. His research interest includes politics, consumption, lifestyles, action, reflexivity, and transformative learning in relation to various transnational environmental and sustainability issues. Boström has studied the social drivers behind unsustainable mass/excess consumption and is currently involved in several research projects on possible counteraction to such drivers among people and movements. Boström is also engaged in theory development and conceptual discussion in environmental sociology. Boström recently published the book *The Social Life of Unsustainable Mass Consumption,* and with Rolf Lidskog, the textbook *Environmental Sociology and Social Transformation—Key Issues.*

Simon R. Bush is Professor and Chair of Environmental Policy at Wageningen University. His research focuses on global environmental governance arrangements for sustainable food, renewable energy, and biodiversity conservation.

Isabela Carmo Cavaco is a Ph.D. student of Environmental Science at the University of São Paulo (USP), where she also obtained her Master's degree on the same subject, and Bachelors on Environmental Management. She has been researching climate justice since 2022, having conducted fieldwork with impacted communities such as traditional, rural, and coastal, focusing on young people. She has international research experience about youth climate movements in the Netherlands. Her Ph.D. is about community experiences regarding territorial planning in preservation areas in São Paulo, Brazil.

Debra J. Davidson is Professor of Environmental Sociology in the Department of Resource Economics and Environmental Sociology at the University of Alberta. Debra began her career at the University of Alberta as Assistant Professor in 1999. Since that time, her teaching and research have been focused on the social impacts of and responses to climate change, with particular attention to energy-society relations. Recent research focused on the role of emotions in social responses to climate change is featured in her latest book, *Feeling Climate Change: How Emotions Govern Our Responses to the*

Climate Emergency (Routledge, 2024). Her work is also featured in several journals, including *Science, Nature, Global Environmental Change, and British Journal of Sociology*. She was Lead Author in Working Group II on the 5th Assessment Report of the Intergovernmental Panel for Climate Change, and is currently a member of the United Nations Environment Program's Expert Foresight Panel. Between 2013 and 2023, she served as Director of Prairie Urban Farm, a community farm at the University of Alberta.

Jennifer E. Givens is an Associate Professor of Sociology at Utah State University. She is an environmental and comparative international sociologist. Her main research areas include environmental and climate change concern, negotiations, mitigation, and adaptation. Specific areas of interest include the carbon intensity of well-being and ecologically unequal exchange relationships.

Philippe Hanna is an Assistant Professor of Spatial Planning and Impact Assessment at the University of Groningen, bringing over two decades of anthropological experience to the field of environmental planning. Based on extensive collaboration with Indigenous Peoples and traditional communities, he advised international development agencies on the cultural adequacy of proposed projects. Bridging academic research and practical application in an interdisciplinary manner, his work provides a comprehensive understanding of environmental and social impact assessment processes and their real-world implications, providing valuable insights into the complexities of social movements and energy projects

Pedro Roberto Jacobi has a Bachelor's degree in Economics and Social Sciences from the University of São Paulo, a Master's degree in Urban and Regional Planning from Harvard University, 1976, and a Ph.D. in Sociology from the University of Sao Paulo, 1986. He is a Senior Full Professor and Research Fellow at the Institute of Energy and Environment of the University of São Paulo, and Full Professor at School of Education/USP (1988–2018). Researcher at the Institute of Advanced Studies at USP. Research Fellow at CLIMARES/USP—Climate Crisis and Disasters Resilience Research Center. Editor of the journal Ambiente & Sociedade/National Association of Graduate Programs and Research in *Environment and Society* (Anppas) since 1997. Since 2011 is President of the Board of Local Governments for Sustainability -ICLEI- South America. His research focuses on urban sustainability, water governance, metropolitan governance and climate change, climate justice, solid waste and sanitation, socio-environmental governance, social learning, and education for sustainability. His last publications are as main editor of Environmental Governance at the São Paulo Macrometropolis (Springer, 2024), co-editor of Towards a Just Climate Change Resilience (Palgrave, 2021), co-editor of Adaptation and Public Policies at the São Paulo macro metropolis: a science-policy approach (Institute of Energy and Environment/ USP, 2020), co-editor of Water Conflicts and Hydrocracy in the Americas—Coalitions, Networks, Policies (IEE/USP & Fapesp, 2018). He also published several articles on Metropolis and Climate Change, Climate Justice, Social Learning and Water Governance.

Sara Karimzadeh is an Assistant Professor of Sociology at Örebro University. She is an established scholar in the sociology of consumption and has contributed to the conceptual and empirical development of research on sufficiency, care, sustainability, and ethical consumption. Most recently, she has developed the concept of "waste delinking" as a mechanism for tackling unsustainable consumption and production, and as a pathway toward degrowth.

C. S. A. Kris van Koppen was Associate Professor Environmental Policy at Wageningen University and Professor of Environmental Education by special appointment at the Utrecht University. He now works as Emeritus Fellow at the Environmental Policy Group of Wageningen University. His research and educational interests center around the sociology of nature, social and organizational learning, and public sociology.

Natalia Landívar is a racialized migrant woman born in Guayaquil, Ecuador, and residing on Treaty 8 territory since 2024. During her earlier studies in Economics at Heidelberg University in Germany, she began collaborating with FIAN International, an engagement that shaped her long-term commitment to human rights and food sovereignty. In 2016, she founded FIAN Ecuador, where she served for a decade, working closely with peasant and small-scale food producers' organizations, as well as with regional and international networks, to advance struggles for land rights, food sovereignty, and broader human rights. From 2014 to 2020, she also served on the Board of Directors of SeedChange, a Canadian non-governmental organization supporting biodiversity, resilient food systems, and the rights of small-scale farmers in Canada and the Global South. Since 2019, she has been a member of the Permanent Committee for Human Rights (CDH) in Guayaquil, an organization that supports low-income communities facing human rights violations. In 2023, she earned her Ph.D. in Natural Resources Management from the University of Manitoba. Her doctoral research built on her Master's degree in Territorial Rural Development from FLACSO–Ecuador and drew on more than a decade of sustained activism alongside peasant associations in her home region, particularly in struggles over land, agrarian justice, and social reproduction. More recently, Natalia began a postdoctoral fellowship at the University of Alberta, where she works as a sessional lecturer and researcher focusing on energy transition and climate justice.

Rolf Lidskog is Professor of Sociology at Örebro University, Sweden. His research interests include environmental sociology, risk sociology, and science and technology studies. Currently, he is investigating the role of expertise in international environmental governance. His latest book is Environmental Sociology and Social Transformation (Routledge, 2024), co-authored with Magnus Boström.

Victor Marchezini is a sociologist at the Brazilian Early Warning Center (Cemaden). He is a professor at the Doctorate Program on Earth System Science, National Institute for Space Research (PGCST/INPE), and at the Graduate Program on Disaster (a joint initiative between ICT/Unesp and Cemaden). He is coordinating the COPE project (Fapesp grant #2022/02891–9).

Henrike Rau is Professor of Social Geography and Sustainability Research at LMU Munich, Germany. She has made internationally recognized contributions to the conceptual and methodological advancement of social scientific and interdisciplinary sustainability research on topics such as domestic energy use, food consumption and mobility practices across the life course.

Juan Saavedra is a social worker and Doctor of Human Sciences. He is an associate professor in the Department of Social Sciences at the University of Bío-Bío (Chile) and a researcher at CETI-UBB. He is currently the lead researcher on the ANID-Chile Fondecyt project No. 1251794.

Göran Sundqvist is professor of science and technology studies (STS) at the University of Gothenburg. His research is on the interplay between science and policy and the role of expert knowledge in climate transition. His most recent book, together with Linda Soneryd, is *Science and Democracy* (Bristol University Press, 2023).

Pedro Henrique Campello Torres is a Professor at São Paulo State University (UNESP) and the University of São Paulo (USP). His research focuses on socio-environmental planning and management, with particular emphasis on environmental inequalities, climate justice, just climate adaptation and coastal cities. He coordinates the LADAMPS—Laboratory for Research on Environmental Inequalities, Climate Change and Planning in Socio-Ecological Systems. Pedro Torres is a Lead Author of the IPCC Seventh Assessment Report (AR7), Working Group II: Impacts, Adaptation and Vulnerability.

Ethemcan Turhan is an assistant professor of environmental planning at the Department of Spatial Planning and Environment, University of Groningen. His research and teaching interests are situated in the broadly defined field of political ecology with empirical attention to environmental conflicts, energy infrastructures, social movements, and climate mobilities. He co-edited "Transforming Socio-Natures in Turkey: Landscapes, State and Environmental Movements" (Routledge, 2019) and "Urban Movements and Climate Change: Loss, Damage and Radical Adaptation" (Amsterdam University Press, 2024).

Steven Yearley is the Professor of the Sociology of Scientific Knowledge at the University of Edinburgh. He has been working on the links between scientific expertise and environmental issues, campaigns, and policies for several decades. His most recent large-scale research project was on industrial decarbonization in the UK and he is currently developing work on greenhouse-gas sinks.

List of Figures

List of Tables

1

Introduction

Debra J. Davidson, Magnus Boström, and Pedro Roberto Jacobi

Since publication of the first edition of this volume (Boström & Davidson, 2018), eruptions of a political nature have dominated our attention. Backstage but no less consequential, manifestation of the impacts of anthropogenic warming and their interactions with other forms of environmental disruption have unfolded at a pace and scale that have sent even climate scientists reeling. Meanwhile, awareness has grown regarding other much quieter but no less consequential environmental concerns: the rate of species extinction escalating to such an extent that it has achieved notoriety as the sixth mass extinction in our earth's history, and the long-term effects of the accumulation of waste, particularly microplastics, in our oceans and our bodies are nothing short of alarming. These events are not independent of each other; they describe a current moment of polycrisis, with crises erupting in multiple inter-dependent systems each of which exacerbate the crisis potential in other systems. What have also become blatantly clear are the stark inequities in the distribution of both accountability and impacts, with those individuals

D. J. Davidson (✉)
University of Alberta, Edmonton, AB, Canada
e-mail: debra.davidson@ualberta.ca

M. Boström
Örebro University, Örebro, Sweden
e-mail: magnus.bostrom@oru.se

P. R. Jacobi
Universidade de São Paulo, Sao Paulo, Brazil
e-mail: prjacobi@usp.br

D. J. Davidson et al. (eds.), *Environment and Society*,
https://doi.org/10.1007/978-3-032-27059-7_1

and organizations bearing the greatest responsibility for producing the crises being the most insulated from (and insured against) the catastrophes that then unfold.

Once relegated to science fiction, the question of existential civilizational survival has become a little too reality TV for comfort, with tips on evacuation planning and surviving the apocalypse now readily found in mainstream media. While the system shocks already experienced can be expected to continue, collective social responses to this polycrisis will have a substantive influence over the nature and extent of future decline and revival. What is now clear is that our potential to minimize decline and maximize revival demands transformation of the institutions defining modern society. Many aware citizens look to news of the latest scientific and technological developments for the drivers of such transformation. In this volume, however, we draw attention to a different set of phenomena entirely, which serve as the drivers of those drivers, as it were. Our institutions, and the decisions made within them, are founded upon assumptions, beliefs, and worldviews which serve as their sources of legitimacy. Those assumptions, beliefs, and worldviews tend to remain hidden within *concepts*: abstract ideas that serve to guide our definitions of problems and ways to address them. Consider for a moment the seemingly innocuous concept of progress, and how it has shaped political and socio-economic histories for the past 300 years.

Concepts serve as important orienting devices for collective action in all its forms. However, concepts will best serve collective projects to confront our current polycrises when they are dynamic: shaped by ongoing feedback, informed by multiple knowledges, experiences and changing historical circumstances. Many of the most powerful concepts, however—like progress—all too often become reified, persisting even when emerging evidence calls into question their validity. Some concepts can serve to inhibit institutional transformation when such transformation is urgently needed, or enable the persistence of inequities which themselves serve to inhibit transformation.

Our attention to social and environmental problems has always emerged through the concepts we embrace to comprehend society's relationship with the natural world. This observation would be of purely academic interest if it weren't for the fact that decisions are made on the basis of those concepts; decisions that have bearing on the well-being, and even survival, of present and future generations. Because these decisions are made less on the basis of 'facts,' and more on the basis of dominant and at times competing interpretations and meanings created and adopted by different societal actors, it behooves all scholars, decision makers, and citizens with an interest in environmental well-being to closely scrutinize these interpretations and meanings and those actors who produce them. This book seeks to invite reflexivity in all spheres of society, including academia, policy, and practice, by scrutinizing a curated set of concepts that we see as having influence—potential or realized—in shaping discourses regarding environment-society relations in ways that can be productive, or constraining of our efforts to pursue just transformation. This scrutiny includes consideration of which concepts are used and how, and what assumptions and premises underpin them.

Applying a critical gaze to the concepts that dominate discussions of environment-society relations allows for deeper reflection not only of their influence over the pursuit of just transformations, but also of their fungibility. Such scrutiny creates openings for invoking certain concepts in new ways, or *re*-conceptualization, which can inform dialogues, policies, and practices. The power to shape knowledge, interpretations, and public debate through the deployment of concepts is and has always been a core weapon in the battle for a more ecologically sustainable and just world. Scholars, decision makers and citizens perceive, understand, explain, and address socio-ecological crises with the knowledge and discourses at their disposal. Concepts not only represent our world, but are part of the shaping of that world (Rau & Fahy, 2013). Institutions such as Bretton Woods, the United Nations, the World Trade Organizations, and the Intergovernmental Panel on Climate Change have been shaped by theories and concepts such as Keynesianism, Human Rights, Neo-classical economics, Sustainable Development and many more (White et al., 2016). Lidskog and Waterton (2016a, p. 308) explain: "Concepts do something with the world. They are navigational (directing our attention), normative (shaping our priorities) and performative (guiding our action)."

Science, policy, and practice share the same concepts to a significant extent, presenting both an opportunity and a challenge in efforts to navigate socio-ecological crises. For instance, ecologists use concepts like resilience to study environment-society relations, but the same or similar concepts appear in the lexicon of urban planning manuals, in ways that only vaguely resemble the concept as it is employed in ecology. To further complicate the matter, researchers are also part of what they study—a situation social scientists call a 'double hermeneutic' (Giddens, 1984)—although this is not often acknowledged. An invitation to scrutinize concepts thus compels reflexivity on the part of the actor; an invitation for those of us who are concerned about justice and environmental well-being to take a good look at the selectivity and partiality of our own analyses.

Concepts used to understand society's relationship with the natural world have been featured in academic discourses for over a Century. These have undergone some notable shifts over the years, with the introduction of new concepts, and the discarding of others. The 1960s saw the emergence of concepts such as Ecology and The Tragedy of the Commons, and powerful metaphors like Silent Spring and Population Bomb. In the 1970s came Limits to Growth, subsequently subjected to heated criticism and debate. These debates focused attention on population growth, pollution, scarcity of natural resources and the role of economy and technology. The key global concept introduced a decade later was Sustainable Development, which countered the Limits to Growth concept with a more optimistic view of economic development, albeit one that attempted to transcend old Enlightenment stress on eternal progress, growth, and prosperity. Rather, development was to be seen as something to be (and which *could* be) balanced with concerns for the environment and social justice. Environmental protection and equity were now to be seen as both current affairs and concerns for future generations. And while environmental scientists began to utilize Elinor Ostrom's Common Pool Resources concept (van Koppen and colleagues take up the commons concept in Chapter 8), the environmentally conscious citizens of the 1990s began to assess the size of their Ecological Footprints and engage in Green Consumption (see Rau and Karimzadeh, Chapter 7 on the concept of Demand). The conceptual treatment of the environment across the academy has been especially dynamic since the turn of the twenty-first Century. Increasingly today, new concepts originating in academia have become part of the lexicon of public debates regarding the environment and the climate emergency, with direct bearing on policy and practice, concepts like resilience, the Anthropocene, planetary boundaries, and tipping points.

Through these concepts, many stories are told about environment-society relations, as problematic, disastrous, catastrophic, or merely inconvenient. Questions arise, however, about how concepts could facilitate learning about environment-society relations in ways that are more insightful than the many simple statements on how 'society' causes too much population, consumption, production, capital, exploitation, extraction, pollution and so on. Concepts narrow our gaze, and thus not surprisingly they tend to polarize, as between neoliberal Promethean optimists and Malthusian 'end times' pessimists (White et al., 2016). In one camp, nature is the problem and society (science, technology, market) the solution. In the other camp, society is the problem and nature the solution.

This book offers appraisal, problematization, and critiques of many contemporary concepts in environmental parlance. Most of our contributors emanate from the environmental social sciences. Despite variations in concepts and approaches, all chapter authors join in a common desire to grapple with how we—scholars, decision makers, citizen-consumers—think and talk about the natural world and our role within it, with an eye toward directing these discussions toward pathways that support transformations toward ecological sustainability and justice. Thus, each of the chapters included will review and discuss a particular concept and related subconcepts in use by researchers and practitioners, including consideration of their limitations and their contributions to our general understanding of environment-society relations, with particular attention to potential distinctions between Global North and Global South interpretations, the implications of their manifestation in the political and public spheres, and specifically their relevance for just transformation.

We have an additional objective, however, in putting together this edited collection. We fully agree with Kasper (2016) that the differentiation between disciplines and sub-disciplines is both an opportunity and a challenge in environmental sciences. It enables specialization, yet the benefits of this specialization become constraints because of the siloed nature of our disciplinary structure. Rather than facilitating shared understanding of society-environment relations, it limits such understanding. In hopes of challenging the limitations imposed by disciplinary silos and facilitating greater transdisciplinary exchange, we purposely chose to focus our attention on concepts that appear to resonate throughout the environmental sciences and beyond. By doing so, we call on all scholars in the environmental sciences and studies to enhance our capacities to engage in cross- and transdisciplinary discussion and debate.

Why Concepts?

The use of concepts is an essential, and inevitable component of research in all disciplines, as well as among actors outside academia. We argue that the employment of any concept must be done so consciously, coinciding with a critical understanding of the underlying implications of that concept: which causal mechanisms are brought to the fore, and which are hidden? What does that concept imply about the relationship within society and of society with the natural world of which it is a part? In a world in which record-breaking world average temperatures, melting ice sheets, biodiversity loss and chemical pollution coincide with mounting military violence, famine, and the disintegration of stability in Western democracies, it behooves us to understand how concepts matter more than ever before. Where we go from here has as much to do with the concepts we use to convey knowledge as has ever been the case before. Which concepts are likely to support climate movement mobilization? Which concepts are likely to support further empowering corporations? Which concepts are likely to support dictatorship, and which democracy? Which concepts provide grounding for the pursuit of just transformations? Scholars, decision makers, environmental experts and citizen-consumers rely upon concepts to understand the infinite and complex phenomena that make up the world around us, and how these phenomena link to the production, distribution, and resolution of socio-ecological problems.

'Concepts' can be flexible and open-ended, and thus able to facilitate interchange across disciplines. Paradoxically, concepts need to 'fixate' something (Hjørland, 2009), while also adapting to various situations (Prinz & Clark, 2004). 'Fixation' is a prerequisite to be able to think, communicate, and act upon something, and may involve including certain elements and excluding others. For example, Sustainable Development excludes unlimited economic growth. However, Sustainable Development at the same time offers interpretive flexibility, enabling different groups and conflicting interests to enter a dialogue (Hajer, 1995). Hjørland (2009) addresses four different sources of conceptualizations. Concepts may derive from systematic observations based on the clustering of similar objects (empiricism), from logical reasoning and theorizing (rationalism), from the historical, cultural, discursive, and social context (historicism), or they may be future-oriented, derived from goals, values, and aspirations (pragmatism). For the purpose of this book, we uphold the assertion that concepts are not only derived from the scientific enterprise but involve broader contexts, including pragmatic policy- and goal-oriented activities.

As with many other fields of inquiry, social science and humanities scholars focused on environment-society relations work with an extensive conceptual plurality that is an indication of the richness and vitality of the field, and provides opportunities for reflexivity, specialization, and synthesis (Lidskog & Waterton, 2016b).

Through the critical analysis of a set of key concepts in environmental studies, throughout this book we draw attention to and interpret problems and solutions, risk and safety, justice and injustice, opportunities and barriers, potential futures and crucial legacies, general patterns, and local variations. A focus on concepts invites several issues with which the contributors to this volume engage:

- Intellectual background and understanding of a core concept in the environmental social sciences. Each chapter summarizes and reviews a particular concept as well as engages in a discussion on how that core concept can contribute to societal efforts to address environmental and climate disruptions, by providing explanation, understanding and critique of the social generation of and responses to those disruptions.
- Critical examination, where relevant, of to whom, for whom, and about whom the concepts speak. All concepts embody frames that resonate with certain audiences more than others.
- The contemporary relevance of each concept to local, regional, and global prospects for the emergence of just transformations in response to the current polycrises we face.
- The multiple ways these concepts manifest in research, politics, and practices in multiple settings, with a particular sensitivity to potential distinctions between Global North and Global South articulations. This includes possible social, cultural, or geo-political 'biases' and 'blinders' inherent in different concepts. Are these concepts globally relevant or are they only germane for a particular region of the world (e.g., Europe)? Or for a particular generation, ethnicity, class, or gender?

Conceptual Traps in the Study of Environment-Society Relations

In addition to discussing the role and importance of core concepts, we find it necessary to emphasize some conceptual traps that scholars ought to have in mind when theorizing and reflecting on environment-society relations. Acknowledging that it is neither necessary nor possible to elaborate on all kinds of common conceptual traps, such as circularity, atomism, or conceptual fuzziness, we encourage readers to consider those that affect how the concepts featured here have evolved historically in sociology and environmental sciences and that appear problematic for the critical-constructive study of society-environment relations today. Indeed, scholars and other actors need to be reflexively aware of how our conceptual framings of reality facilitate or inhibit social change, including how basic ontological and epistemological assumptions may affect these framings and create vulnerabilities to certain conceptual traps, or combinations thereof.

Reductionism

Concepts are needed that can capture environment-society relations, their implications for just transformation, and in addition enable comprehension of our interconnected world, both spatially and temporally. White and colleagues (2016, p. 17) understand social reductionism "as the tendency to underplay the importance of material forces on society, the assumption that 'culture,' 'history,' 'society,' or 'discourse' trumps everything." Since Emile Durkheim formulated his methodological rules for sociology more than 100 years ago, sociologists everywhere have been taught to think that society exists (*sui generis*) and to explain social facts by other social facts. The avoidance of consideration for the interaction or entanglement of society with biophysical conditions as well as bringing in such conditions in efforts to explain social dimensions has not just been a tendency but an explicit disciplinary strategy. Several chapters in this book express the importance of countering this tendency, and explicitly discuss material dimensions.

On the other hand, *ecological reductionism* is systemic in the environmental sciences. In many environmental conceptualizations, the ecological domain is treated as something objectively fixed and given, something to which the social domain must be oriented. Perhaps more often the social is treated as a single variable, factor or pillar, or a limited list of elements, or simply a black box (cf. Brulle & Dunlap, 2015, p. 15). In other words, the social dimension is reduced to (just) one measurable and manageable aspect, such as 'population' (to be controlled), 'population growth' (to be stopped and potentially reversed), 'attitudes' (to be changed), or 'the public' (to be educated). In their discussion about the 'Anthropocene,' Lidskog and Waterton (2016b) warn that this concept could result in naturalization, and inattentiveness to issues of power, intersectionality, politics, and injustice (see also Boström & Lidskog, 2024 about the meaning of the social).

Dualism vs Hybrids

Reductionism has affinities with dualist thinking. As noted by White et al. (2016), the irresolvability of many of the debates on society-nature relations can be attributed to their binary underpinnings. Some of the more fundamental are Society vs. Nature, Material vs. Culture, Body vs. Mind, and Human vs. Non-human. Very significant are also distinctions such as Malthusian pessimism vs. Promethean optimism (Dryzek, 2012), Growth/Degrowth, and political/geographical divisions, such as North/South, East/West, Developed/Developing, Democratic/Totalitarian just to list a few.

White and his colleagues introduce 'hybridity' as a way to overcome dualist and binary thinking, with reference to theorists such as Bruno Latour, Donna Haraway, Ulrich Beck, David Harvey, and others. These theorists highlight a more relational and entangled view of the society-nature relation (and similar distinctions), and today scholars have at their disposal many new and fruitful conceptualizations of the entangled relationships between individuals, organizations, and institutions.

However, contemplations of hybridity have themselves been criticized (Dunlap, 2015; Murphy, 2016). Murphy (2016), for example, argues for the need to conceptually disentangle what has become empirically entangled. Only in this way can we study the interactions between 'social constructions' and 'nature's construction,' he argues, and be in a position to study variations of empirical hybrid manifestations. "The concept of co-constructed hybrids is an indiscriminate grab-all notion that must not be allowed to degenerate into conflating two referents just because they are found together empirically, thereby obscuring significant differences and relationships. It lacks specificity and raises more questions than it answers" (Murphy, 2016, p. 338). White and colleagues also urge scholars to 'relativize' what is 'human' in much hybrid thinking: "Without the capacity to think through human/non-human points of *continuity and difference*, we have no social science, nor do we have any capacity to respond coherently, ethically and politically to the socio-environmental dilemmas of our time" (2016, p. 35). Environmental social scientists thus face a challenge to bring conceptual order to a messy, entangled and changing empirical reality, which some postmodern social scientists call 'hyper-reality.' Concepts are needed that can sufficiently simplify empirical reality to make it comprehensible, but not to the extent that it becomes reductionist.

Rigid vs. Elastic Views of Society

The theorizing of environment-society relations requires concepts that account for (the potential of) agency and social change. In the social sciences, concepts such as 'transition' and 'collective action' invoke crucial questions about opportunities and barriers for change. Chapters in this book engage in concepts that all have relevance for change-oriented work. Equally important are theories and concepts that focus on why change might not happen. In these theorizing efforts, however, there is a risk of over-emphasizing the rigidity of social structures, stability, social order, inertia, and path-dependencies, while overlooking processes that imagine, conceptualize, and facilitate change.

Yet researchers can err too far on the other end of the spectrum as well. As an example, White et al. (2016) criticize Latour's Actor-Network-Theory for its over-elastic view of society. "Latour's tendency to view all consolidated social-ecological-technological forms as dissolving into fluid networks or assemblages would seem, as its core, to underestimate *the sticky obduracy of certain key social processes and institutional forms* that have shaped modernity, notably capital and processes of labor, nation-state boundaries and political interests, patriarchy, racism, imperialism, and so on" (White et al., 2016, p. 133). This implies a rather shallow view of society that can easily be constructed, and then torn apart and reconstructed in another way. As will be shown throughout this collection, such theories fail to conceptualize rigid structures, deeply embedded relations, and path-dependent historical processes.

Environmental scientists also need to be reminded to avoid adopting an over-elastic view of society, such as those mirrored in calls for *urgent* and *dramatic* societal transformation to save humanity from self-destruction. In these debates about solutions, words such as 'innovation,' 'green technology,' 'regulation' and 'decision-maker' frequently feature, often in an uncritical way. The use of such concepts will certainly be necessary if aspirations to 'rebuild' and 'transform' societies are to succeed. However, calls for change without due consideration for the more resonant aspects of society, like culture, are likely to remain ineffective or even fuel resistance, regardless of how urgent the transformation is.

Greenwash vs. Hyper-criticism

Does a particular concept evoke problem-solving or problem-maintaining? Murphy (2016) argues that problem-solving conceptualizations are awkward, even dangerous, if problems are in fact escalating. A 'false-positive' impression helps actors legitimize problematic practices. In environmental studies, the word greenwashing is frequently used to critique how concepts such as 'sustainability' and 'responsibility' are invoked in policy and practice to legitimize questionable practices. However, a 'false-negative' impression may lead to other problems such as resignation and fatalism. Hyper criticism, an a priori rejection akin to social discrimination, can be implied when someone criticizes something or someone else for whatever they think or do.

This pendulum between greenwash and hyper-criticism has also been observed in the heated debates that have taken place within environmental sociology between two core bodies of theory: *treadmill of production* (see e.g., Schnaiberg et al., 2002), and *ecological modernization* (see e.g., Mol, 2000).

While the former emphasizes the apparent inevitability of environmental destruction caused by capitalism's expansionist tendencies, the latter instead focuses on the creation of an ecological rationality in and through that very capitalist system. The former places blame on a growth coalition including capital, the state and labor; the latter considers how and in which contexts those same institutions can be reformed. These two competing theories and their associated conceptual underpinnings have fueled research that draws markedly different conclusions. A new technology, for example, may be interpreted as a sign of an ecological rationality, or a form of 'greenwashing' that effectively conceals the fact that environmental degradation has simply been relocated to developing countries with cheap labor and poor environmental regulation. We are not suggesting that adherents of either theory *necessarily* espouse greenwashing or hyper-criticism, but rather that strict adherence to any single set of theoretical premises constitutes a form of conceptual trap, leading to the tendency to favor the selection of research inquiries to support the theory, rather than the other way around.

Grand Theory vs. Narration

Finally, the warnings posed by C. Wright Mills and Robert Merton many decades ago retain their relevance today. As stated by Mills in *The Sociological Imagination*, grand theorizing entails "a level of thinking so general that its practitioners cannot logically get down to observation" (Mills, 1957, p. 33). For the grand theorists, concepts become 'fetishized,' argues Mills, and such theories tend to be associated with monocausalism: attributing outcomes to a single cause. Some theories, for example, may postulate that social change is driven entirely by capitalism (e.g., Marxism), the division of labor/social differentiation (A. Smith/E. Durkheim), or a process of rationalization (M. Weber). Such tendency usually involves theory-supporting rather than theory-testing. This tendency can still be observed across the environmental sciences, such as the debate between ecological modernization and treadmill of production discussed above. Yet others grapple with the proliferation of overly descriptive empirical studies in environmental social sciences, including both single case qualitative studies, and quantitative analyses of various statistics and survey data, which offer few opportunities for conceptual development.

Finally, 'theories of the middle range' (Merton, 1949) engage a plurality of core theories and concepts. It is this middle way of theorizing that engages what Mills (1957) calls the 'sociological imagination', which is a quality of mind, expressing the capacity to shift from one perspective to another, to see that things could be different, and views contemporary events as part of history and part of creating history. This middle way also embraces reflexivity, which itself has appeared as a core concept in environmental sociology. The topic of denial, which started our discussion of this introductory chapter, has been conceptualized as a struggle between reflexivity[1] and anti-reflexivity. The risks and the uncertainties of contemporary times and the destabilization of authorities of modern societies could lead to reflexivity—understood as a critical-constructive mind-set—but could very well also lead to the opposite of reflexivity: denialism, fundamentalist ideologies, nationalism, and xenophobia.[2] While acknowledging that reflexivity is not the antidote to every problem (see Boström et al., 2017, 2024), it is nevertheless a necessary ingredient for both our sociological imagination and our ability to navigate between the conceptual traps discussed previously.

Moving from Polycrisis to Just Transformation

This second edition specifically focuses upon a set of concepts that we believe have particular relevance to collective efforts to engage in just transitions. Some of the concepts included here, or earlier variants of them, were also featured in the first edition of this volume. Others are new. Importantly, the list is by no means exhaustive. What is new in this edition is a specific call to our authors to consider how the deployment of these concepts in academic, policy and public dialogues may inform, or provide enabling conditions for, just transformation, increasingly recognized by environmental studies scholars, the editors included, as crucial to our ability to navigate our current polycrises. However, as we know from history, rapid and disruptive shifts to social systems can induce enormous harm, and therefore intentional efforts to support institutional transformation must be attentive to equity and justice. The prioritization of justice in our collective transformation projects can even be seen as a crucial prerequisite to the long-term success of those efforts. Blyth et al. (2019) remind us about "the dark side of transformation," and

[1] The concept is particularly associated with the theories of risk society and reflexive modernization, developed by Ulrich Beck and Anthony Giddens in the 1980s. It is associated with a rather optimistic account of the propensity of individuals, experts and organizations to, based on the experiences of escalating risks and hazards, reflect on how existing practices reproduce problems, and based on this develop new practices (see Boström et al., 2017 for a review). Even if risks and environmental problems escalate, the theory of reflexive modernization holds that there is an increasing propensity for reflexivity due to historical processes of individualization and the undermining of traditional authorities and structures (e.g., the state, church, science, gender roles).

[2] These potential tendencies have been analyzed by a number of scholars including Beck (2009) and McCright and Dunlap (2010).

discuss five risks associated with the current dominant discourse on sustainability transformation: (a) the discourse attributes responsibility to vulnerable parties, e.g., to citizens and communities to build their own resilience; (b) it is used to justify-business-as-usual, with, for example, a narrow focus on green growth and technological change; (c) it fails to consider winners and losers; (d) it loses sight of the possibility of resistance; and (e) insufficient recognition of the role of politics and power. Their warnings suggest the importance of systematically connecting justice to our discussions of transformation.

Both *justice* and *transformation*, however, are themselves concepts that demand scrutiny. The concept of justice has of course been subject to extensive critical scrutiny and academic debate, including in the environmental social sciences. The first edition of this volume featured a chapter on justice, and it is featured in this volume as well. Transformation is often used interchangeably with the term 'transition.' In a recent comprehensive literature review around the concept of 'just transition,' Wibeck et al. (2026) show that attention to the concept has grown rapidly in recent years in both policy and research, still with the majority of research focused on Global North. The authors see just transition as an 'emblematic' concept that acts as rhetorical anchors or reference points and with potential to bridge meanings and actors. "Such concepts can serve to mobilize support, legitimize political action, and unite coalitions by condensing a variety of meanings, values, emotions, and recommendations" (ibid., p. 2). However, 'transition' is often used to refer to the passage between states, as from A to B, within a specific domain, like shifting from one technology to another, for example, or from one fuel to another. Transformation implies a much broader set of changes, and along with it a much broader set of implications (economic, political, cultural, infrastructural, etc.). Wibeck et al. (2026) observe that 'transformation' is often used interchangeably with 'transition,' although transformation generally denotes a more comprehensive view of society and social change. The concept of transformation has been employed in ecology to describe a shift in the structure and function of an ecosystem, in contrast to resilience, which describes a system that responds to shocks while retaining the basic structure and function of that system. When applied to social systems, Boström and Lidskog (2024) describe five facets of transformation, reflecting the encompassing character of transformation events: (a) change of inner lives; (b) change in and through social relations; (c) change in socio-material entities and practices; (d) change toward justice and equality; (e) institutional change.

Transformation has also been employed in research focused on climate change impacts and adaptation, understood as "profound and enduring nonlinear systematic changes, typically involving social, cultural, technological, political, economic, and/or environmental processes" (Linnér & Wibeck, 2019, p. 4), with common appearances in scientific reports produced by the Intergovernmental Panel for Climate Change (IPCC) and the Intergovernmental Science-Policy Platform on Biodiversity and Ecosystem Services (IPBES), often juxtaposed to 'incremental adaptation.' More recently, the concept has found its way into certain policy texts, most notably the Paris Agreement on Climate released by the UNFCCC in 2015; an inclusion which no doubt increased its popularity enormously (Linnér & Wibeck, 2019). In contrast to incremental adaptation, Linnér and Wibeck (2019) rightly note transformation's inherently radical character, posing as it does threats to incumbent institutional interests.

While many across the environmental social sciences and humanities may agree the need for transformation in principle, there is little agreement on how such transformations are likely to occur, or even how we are able to observe their occurrence, not to mention whether such transformations will abide by the principles of justice. What is to be transformed, by whom, how such transformations ought to be governed, or if they can be governed, are matters of serious debate. But there is more: Is everything to be transformed? What is essential to keep as is? Are we even capable of such 'managed transformations' given the emergent effects of disruptions in complex systems?

The notion of transformation relates to the action-potential implicated by a particular concept. Many concepts shape action (understood as performativity), intentionally or unintentionally. Does the concept generate openings for vision, imagination, politics, innovation, institutional design, for taking action, for identifying transition pathways? Who are the implied change-makers? Are these agents human? Will artificial intelligence lead the transformation? Or unintended consequences like catastrophes?

Introducing the Chapters

Contributors to this volume, all working within the field of environmental social science, provide critical insights into the questions that we have raised in this chapter. Each chapter grapples with problems and assumptions associated with how we think and talk about society-environmental relations, with an eye toward directing these discussions toward constructive responses to environmental problems and the need for just transformation.

For obvious reasons, a volume like this cannot and should not cover all concepts that appear in environmental sciences and policy debates,[3] but can offer a deeper analysis of some selected significant concepts that are commonly applied in environmental sciences and debate. We hope that the reader will be encouraged to scrutinize other significant concepts in the field with perspectives and approaches similar to those developed by the contributors in this volume. It is our intention to represent a wide variety of core concepts, some with origins in sociology or social sciences more broadly (intersectionality, commons), and others that originated in the natural and other sciences but which have been taken up by social scientists (metabolism, demand), and yet other concepts that are core to the scientific enterprise itself (knowledge). Finally, there are concepts that originate in policy discourse, but which have gradually entered into scientific discourse: key examples are the concepts of justice, colonialism, inequality, and disaster. Concepts dealing with temporal (futures) and spatial dimensions often have more heterogeneous origins involving a broad array of disciplines and discourses, as well as concepts like knowledge and catastrophe. Here follows a brief introduction of each.

Chapter 2 attends to the second concept defining our core objective: justice. After laying out the four dimensions of justice, Isabela Cavaco Ethemcan Turhan, Philippe Hanna and Pedro Henrique Campello Torres, all writing from different regions in South America, highlight the distinct contexts of climate justice in the Global North and South, with case studies of both injustice and the social movements that have emerged to confront them, thereby linking the concept of justice to social movement theories. Chapter 3 follows with a conceptual inquiry into those conditions that presuppose calls for justice, namely inequality. In this chapter, Jennifer Givens and Mufti Ahmed begin by highlighting the ambiguity embedded in this concept, introducing readers to several different lenses within and adjacent to environmental sociology which approach inequality in very different ways. They draw attention to important companion concepts that attend to the ways inequality manifests at the global scale, including ecologically unequal exchange and ecological debt. Importantly, these authors highlight empirical correlations that counter the presumption that environmental impact is a necessary cost to developing our economies to improve well-being. To the contrary, higher levels of inequality correlate with higher levels of carbon intensity and environmental degradation. Providing a segue to the next chapter, Givens and Ahmed note that economic inequality is by no

[3] For example, The Elgar Encyclopedia of Environmental Sociology (Overdevest, 2024) reviews 99 different concept applied in environmental sociology to study environment-society relations.

means the only form of inequality that matters to just transformations: so too do ethnic and gender inequality, suggestive of linkages between patriarchy/sexism, racism, and environmental impact.

Chapter 4 picks up on these threads introducing an important additional conceptual articulation of both justice and inequality, which has enjoyed a relatively shorter history in the environmental social sciences, but has quickly engendered substantive advances in these fields: intersectionality. Ana Terra Amorim-Maia and Isabelle Anguelovski build a conceptual framework for intersectional justice in the context of just transformations, organized around five core principles. With reference to Barcelona's multi-tiered climate action planning as a case study, the authors illustrate both the challenges and the creativity associated with attempts to develop intersectionality-informed urban climate resilience planning. As noted by these authors, "intersectionality inherently involves analyzing power relations and structures of subordination, while encouraging collective responses to social injustices and fostering activism."

We then move in Chapter 5 to the twin concepts of disaster and catastrophe. How do prevailing conceptual understandings of disaster and catastrophe hold up to scrutiny as we consider them from the perspective of justice, and in the context of a climate emergency? Victor Marchezini and Juan Saavedra not only unpack these concepts in the context of our climate emergency, but they also offer a close examination of disaster in a Global South context: bringing in case studies of several extreme events affecting Indigenous and non-Indigenous communities across Brazil and Chile, highlighting the important contributions of Latin American scholars to the study of disaster in the process. According to these authors, the prospect of just transformation will necessarily require a reparative, holistic, gendered, and intersectional approach to disaster.

In Chapter 6, we shift our gaze to broader conceptual understandings of nature-society inter-relations, revisiting metabolism, a concept with a long intellectual history in the environmental social sciences and in other corners of the academy as well. Debra Davidson and Steven Yearley explain how the concept of metabolism draws our attention to three primary necessities of socio-ecological systems, whether that system is a family farm or the global economy: access to adequate inputs to support the system, the presence of mechanisms within that system to process and distribute those inputs to where they are needed, and the need to avoid overexploitation of the environment's capacity to absorb wastes. Because just transformation depends upon a foundation of healthy metabolic relations in socio-ecological systems, the concept of metabolism is necessarily central to the design and evaluation of emerging just transformation projects. With growing concerns regarding the ecological sustainability of our economies, renewed interest in forging alternative economic pathways that re-establish healthy metabolic relations in socio-ecological systems has emerged, and the value of the concept of metabolism is once again evident in informing these intellectual and policy pursuits. This concept may simultaneously offer valuable potential to provide a new frame of reference for identifying essential needs to the sustainability of collective action pursuits.

Sara Karimzadeh and Henrike Rau then explore Demand in Chapter 7, that phenomenon giving rise to one of the key sources of metabolic disruption, our ecological footprints. Noting the fallacies of green growth and decoupling discourses, the authors acknowledge the urgent need for demand reduction, yet the integral relationship between demand and well-being complicates matters considerably. According to these authors, "demand is not merely a reflection of consumer preferences but also a pivotal driver of resource allocation and environmental impact that reflects deep-seated social and cultural conditions." They go on to compare consumption practices in two very different contexts, Germany and Iran, highlighting the extent to which prevailing individualized approaches to study demand overlook the importance of culture and cultural difference when discussing ways to reduce demand/consumption.

In Chapter 8, Kris van Koppen, José Barrena and Simon Bush make the case that just transformation requires critically examining space, place, access, and ownership of planetary elements, bringing the concept of the commons to the forefront. The concept of commons helps us to rethink relations of property, democratic control, and use of material and immaterial resources. The commons concept necessarily invites critical consideration of our spatial frames of reference, from nations to socio-ecological systems to planetary systems, and by extension the 'fit' of institutions operating at those scales to oversee the governance of those commons. In this chapter, van Koppen, Barrena and Bush also introduce 'commoning,' as a practice of resistance to privatization and commodification/marketization of nature, in recognition of the many non-market values of the natural world, seeking to build relationships with nature, rather than ownership and control over it.

Natalia Landívar then turns in Chapter 9 to colonialism, a concept whose rise to prominence in the environmental social sciences is long overdue. The 500-year history of colonialism and its contemporary manifestations have suppressed the lives, livelihoods, and knowledge systems of indigenous peoples around the globe. That history holds substantive implications for socio-ecological relations generally, and for collective efforts to pursue just transitions. Colonialism has served as a constitutive mechanism of global capitalist expansion and thus a fundamental driver of environmental and climate disruption. As such, attending to the climate crisis necessarily implicates colonialism, and no transition could be considered just without a full accounting for the harms of colonialism, as well as decolonization, including land repatriation, and reconciliation with surviving indigenous peoples. This is no one-way process, however; decolonization also uplifts knowledge, skills and value sets held by indigenous peoples that can substantively enhance our collective transformation goals and visions. As Landívar makes clear, however, this project has only just begun. Landívar proposes a conceptual framework for decolonial social transformation. By way of illustration, this chapter articulates the six inter-related analytical dimensions of this framework in the context of energy transition.

Rolf Lidskog and Göran Sundqvist then turn our gaze in Chapter 10 toward knowledge, a concept that is most integral to our expressed interest in concepts. All environmental problems, even their identification *as* environmental problems, depend upon knowledge. But is scientific knowledge alone sufficient for effective decisionmaking? The widening gap between knowledge and action suggests otherwise. Few concepts are taken for granted quite to the degree that knowledge is. However, the pursuit of just transformations demands critical examination of knowledge, not as a set of facts but as a concept, one fraught with deeply contested meanings. As the field of sociology of knowledge makes clear, knowledge is not separate and autonomous from society; rather, it is deeply embedded in society. Indeed, scientism—the belief that science is the only valid way to gain knowledge about the world—marginalizes other forms of knowledge and dismisses non-scientific perspectives as irrational or biased. Yet, these authors warn us, the answer is not to ride the wave of post-truthism observed today. Crucially, the project of just transformation demands recognition that science and politics are intertwined. In our 11th and concluding chapter, Davidson, Boström, and Jacobi revisit the project of just transformation in light of the 10 concepts reviewed in this volume.

References

Beck, U. (2009). *World at risk*. Polity.

Blyth, J., Silver, J., Evans, L., & Brown, K. (2019). The dark side of transformation: Latent risks in contemporary sustainability discourse. *Antipode, 50*(5), 1206–1223.

Boström, M., Berg, M., & Lidskog, R. (2024). Reflexivity and anti-reflexivity. In C. Overdevest (Ed.), *Elgar encyclopedia of environmental sociology* (pp. 474–480). Edward Elgar.
Boström, M., & Davidson, D. (Eds.). (2018). *Environment and society: Concepts and challenges*. Palgrave Macmillan.
Boström, M., & Lidskog, R. (2024). *Environmental sociology and social transformation: Key issues*. Routledge.
Boström, M., Lidskog, R., & Uggla, Y. (2017). A reflexive look at reflexivity in environmental sociology. *Environmental Sociology, 3*(1), 6–16.
Brulle, R. J., & Dunlap, R. E. (2015). Sociology and global climate change. In R. E. Dunlap & R. J. Brulle (Eds.), *Climate change and society: Sociological perspectives*. Oxford University Press.
Dryzek, J. S. (2012). *The politics of the earth: Environmental discourses*. Oxford University Press.
Dunlap, R. E. (2015). Environmental sociology. In *International encyclopedia of the social & behavioral sciences* (2nd ed., Vol. 7, pp. 796–803). Elsevier.
Giddens, A. (1984). *The constitution of society*. Polity.
Hajer, M. A. (1995). *The politics of environmental discourse: Ecological modernization and the policy process*. Clarendon Press.
Hjørland, B. (2009). Concept theory. *Journal of the Association for Information Science and Technology, 60*(8), 1519–1536.
Kasper, D. (2016). Re-conceptualizing (environmental) sociology. *Environmental Sociology, 2*(4), 322–332.
Lidskog, R., & Waterton, C. (2016a). Conceptual innovation in environmental sociology. *Environmental Sociology, 2*(4), 307–311.
Lidskog, R., & Waterton, C. (2016b). Anthropocene: A cautious welcome from environmental sociology? *Environmental Sociology, 2*(4), 395–406.
Linnér, B.-O., & Wibeck, V. (2019). *Sustainability transformations: Agents and drivers across society*. Cambridge University Press.
McCright, A. M., & Dunlap, R. E. (2010). Anti-reflexivity: The American conservative movement's success in undermining climate science and policy. *Theory, Culture & Society, 27*, 100–133.
Merton, R. K. (1967[1949]). On sociological theories of the middle range. In *On theoretical sociology* (pp. 39–73). Free Press. (Original work published 1949).
Mills, C. W. (2000[1959]). *The Sociological Imagination*. Oxford University Press.
Mol, A. P. J. (2000). Ecological modernization theory in debate: A review. *Environmental Politics, 9*(1), 17–49.
Murphy, R. (2016). Conceptual lenses to bring into focus the blurred and unpack the entangled. *Environmental Sociology, 2*(4), 333–345.
Overdevest, C. (Ed.). (2024). *Elgar encyclopedia of environmental sociology*. Edward Elgar.
Prinz, J., & Clark, A. (2004). Putting concepts to work: Some thoughts for the twenty-first century. *Mind & Language, 19*(1), 57–69.

Rau, H., & Fahy, F. (2013). Introduction: Sustainability research in the social sciences—Concepts, methodologies and the challenge of interdisciplinarity. In F. Fahy & H. Rau (Eds.), *Methods of sustainability research in the social sciences* (pp. 3–24). Sage.

Schnaiberg, A., Pellow, D. N., & Weinberg, A. (2002). The treadmill of production and the environmental state. In A. P. J. Mol & F. H. Buttel (Eds.), *The environmental state under pressure* (Research in Social Problems and Public Policy, Vol. 10, pp. 15–32). JAI Press.

White, D. F., Rudy, A. P., & Gareau, B. J. (2016). *Environments, natures and social theory*. Palgrave Macmillan.

Wibeck, V., Jaku, A., Lee, S., & Linnér, B.-O. (2026). The widening scope of just transition research: A review of an emblematic concept. *Environmental Policy and Governance*. https://doi.org/10.1002/eet.70040

2

Climate Justice

Isabela Carmo Cavaco, Ethemcan Turhan, Philippe Hanna, and Pedro Henrique Campello Torres

Introduction

Climate change is not only an environmental crisis but also a question of justice, and societal efforts to address climate change thus need to attend to the justice implications, including critical examination of the concept of justice itself. The uneven distribution of risks and responsibilities means that some groups, regions, and generations carry far heavier burdens than others. This chapter explores climate justice as a framework for understanding these

I. C. Cavaco (✉)
Postgraduate Programme of Environmental Science, University of São Paulo, São Paulo, Brazil
e-mail: isabcvc@usp.br

E. Turhan · P. Hanna
Department of Spatial Planning and Environment, University of Groningen, Groningen, The Netherlands
e-mail: e.turhan@rug.nl

P. Hanna
e-mail: p.hanna@rug.nl

P. H. C. Torres
Department of Biological and Environmental Sciences, São Paulo State University, São Vicente, Brazil
e-mail: ph.torres@unesp.br

D. J. Davidson et al. (eds.), *Environment and Society*,
https://doi.org/10.1007/978-3-032-27059-7_2

inequalities, with a particular focus on the contrasts and connections between the Global South and the Global North.

Climate justice discussions emerged in the global North and global South at different times. While early usages of the term can be found in Agarwal and Narain's (1991) work from India, the term popularized in the global North after the publication of 1999 report *Greenhouse Gangsters vs. Climate Justice* by Corpwatch and the first International Forum on Climate Justice held in The Hague in 2009. However, as noted by Arruda Filho et al. (2024), it faced significant challenges in gaining more visibility in the Global South. Despite this, the climate justice movement has expanded in diverse ways across both regions. Key events for the movement took place in the global South, such as the establishment of the Climate Justice Principles in 2002 in Johannesburg, and the World People's Conference on Climate Change and the Rights of Mother Earth held in Cochabamba in 2010.

Thinking Through Climate Justice

Justice is a central concept for understanding climate change and urban transformations. In this chapter, we will develop a conceptual framework consisting of multiple dimensions described further below, which will serve as a lens to interpret the comparative cases that follow, connecting climate impacts with broader issues of social equity, governance, and political ecology, including two instances of youth climate activism, one in Brazil and the other in the Netherlands—two countries with significant interest in climate action—aiming to reveal similarities, differences, and the conceptual relationships emerging from selected texts. As we will see, climate justice encompasses various dimensions, including gender, ethnicity, income, and the realities of indigenous, fishing, rural and other communities. Youth played and continue to play a central role in climate justice movements, while also being a social group profoundly impacted by climate change, as evidenced by the growing prevalence of eco-anxiety and climate anxiety among youth in particular (Börner, 2023; Ojala et al., 2021).

In the Global North, debates often emphasize mitigation commitments, carbon markets, and technological solutions. In the Global South, the central issues are more directly linked to survival: access to water, housing, energy, and protection from climate extremes. These different contexts reveal how climate justice must address both global structural inequalities and local lived experiences. By focusing on justice, we highlight three guiding questions:

1. Who is most exposed to climate risks, and why? (distributive justice)
2. Who participates in decisions about adaptation and mitigation? (procedural justice)
3. Whose voices, histories, and rights are recognized—or ignored—in climate debates? (recognition justice)

The chapter focuses on the multiple dimensions of justice and climate justice, as well as youth activism for climate justice. Our theoretical framework is grounded in critical climate justice (Sultana, 2021), alongside other global South and Brazilian perspectives (Talukdar, 2022; Torres et al., 2021), and includes insights from the Netherlands (Van Der Loeff, 2022; Wahlström et al., 2019; Zantvoort, 2023). As such, the chapter contributes to the development of the climate justice scholarship and readership by characterizing the profiles of youth activists within this movement and exploring its internal contradictions.

The case studies and examples presented throughout the chapter show how justice is contested, negotiated, and redefined. Our goal is not only to inform but also to provide readers with tools to critically evaluate climate policies, activism, and everyday practices. We encourage readers to see justice not as an abstract principle, but as a living, political struggle that shapes both local communities and international negotiations.

Justice as the Core Concept

Climate justice has emerged as a central framework for understanding how climate change intersects with inequality, vulnerability, and power. Rather than treating impacts as uniform or adaptation as a neutral process, climate justice emphasizes that climate risks and responses are always unevenly distributed. To bridge political and legal theories of justice with the concept of climate justice, we draw on authors who explore this intersection, particularly Sultana (2021), whose critical climate justice framework serves as a guiding lens. Sultana emphasizes that critical climate justice is an ongoing, collaborative process of unlearning and refining the concept by integrating diverse perspectives.

Rooted in feminist theory, Sultana's approach resists reductionism and highlights the interconnections between theories and practical applications. One such example is the loss and damage fund, which functions as both corrective climate justice—by holding polluters accountable and compensating affected communities—and distributive climate justice, as it redistributes resources from wealthy nations to structurally disadvantaged ones.

Similarly, Amorim-Maia et al. (2022) present an intersectional perspective on climate justice, aligning with Sultana's emphasis on overlapping vulnerabilities. However, while Amorim-Maia et al. (2022) focus on operationalizing climate justice within urban planning and adaptation, Sultana (2021) examines broader social relationships, advocating for a redistribution of political power and new alliances between the global North and South.

Additionally, Kashwan (2021) defines climate justice as a means of "recognizing and correcting the unequal distribution of the costs and burdens of climate change and climate responses of various types" (p. 2). His work highlights three key dimensions discussed further below—distributive, procedural, and recognition justice—and traces the historical roots of climate injustices, arguing that acknowledging these origins is essential for meaningful change. Often, discussions focus on climate impacts and solutions without addressing the exploitative structures driving the crisis. For instance, Kashwan (2021) illustrates how market-based solutions such as carbon taxes and the expansion of renewable energy industries, despite their environmental benefits, can intensify extractivism and disproportionately harm marginalized communities.

The Dimensions of Environmental and Climate Justice

Environmental and climate justice are multi-dimensional concepts that help us understand not only the outcomes of environmental policies but also the processes through which decisions are made and whose voices are included. Justice scholars have articulated three dimensions of environmental and climate justice as follows (Adger et al., 2006; Schlosberg, 2007; Sultana, 2021):

1. Distributive justice—Who benefits and who bears the costs of climate risks and adaptation policies?
2. Procedural justice—Who participates in decision-making and whose voices are marginalized?
3. Recognition justice—Which groups and identities are made visible or rendered invisible in climate debates and planning?

Schlosberg's (2007) categorization of environmental justice introduces the historical discussions of justice and its transition into climate justice (Schlosberg & Collins, 2014). Distributive justice concerns the fair allocation of environmental benefits and burdens, ensuring that vulnerable groups are not disproportionately affected. Procedural justice emphasizes equitable participation in environmental decision-making, ensuring that all communities have a voice. Recognition justice is about respecting the identities, cultures, and ways of life of different groups, ensuring that their knowledge and rights are considered in environmental policies. Of particular relevance to climate justice is a fourth dimension: intergenerational justice. Following Adger et al. (2006), Schlosberg (2007), and Sultana (2021), we employ a framework that includes at least these four key dimensions, all of which are relevant for comparing North and South contexts. Some approaches (Juhola et al., 2022), as we will explore later, also include restorative justice.

While Schlosberg's (2007) work provides a solid theoretical foundation, we seek to understand climate justice from the definitions and perceptions of the youth activists studied, acknowledging that these actors use multiple conceptions simultaneously. By connecting these concepts with the perspectives of activists, we explore how different conceptions of justice interact and transform in the context of contemporary climate activism. Therefore, rather than simply recognizing the plurality of ideal theories, it is crucial to understand how these frameworks can guide concrete policies that can transform real societies closer to fairer models, ensuring that justice does not remain an abstract concept but translates into effective institutional and social change.

Distributive Justice

Distributive justice refers to how environmental benefits and burdens are shared across populations. In the Global South, cities face compounded risks from heat, flooding, and landslides that disproportionately affect low-income, racialized groups living in precarious housing. By contrast, in many Northern cities, while infrastructure often buffers the most severe risks, distributive inequalities still emerge—for example, in the siting of hazardous industries or the unequal access to green infrastructure in European or North American metropolitan areas. The comparison shows that while the scale and intensity of risks may differ, distributive injustices persist across contexts.

Procedural Justice

Procedural justice focuses on fair decision-making, ensuring inclusive and transparent participation (Kuehn, 2000; Schlosberg, 2007). In the South, residents of vulnerable neighborhoods often lack meaningful participation in climate planning, as illustrated by limited consultation in the development of local Climate Action Plans in coastal Brazil. In Northern contexts, procedures are often more institutionalized, yet participation can remain tokenistic, with marginalized voices still sidelined. This highlights that procedural justice is not only about formal opportunities to participate but also about the power to influence outcomes. Barriers such as inaccessible public hearings or withheld information create procedural injustices. Recognition justice, as Schlosberg (2007) argues, is fundamental to distributive justice, as social and political exclusion prevents equitable access to rights and resources. Fraser (2007) expands this, emphasizing that justice involves not only material distribution but also cultural recognition and political representation. She identifies three interlinked dimensions—economic structure (distribution), cultural hierarchies (recognition), and political participation (representation). Without recognition, marginalized groups remain excluded, reinforcing systemic inequalities.

Furthermore, the Bali Principles of Climate Justice, developed just before the 2002 World Summit on Sustainable Development (Earth Summit) by an international coalition of social movements (including Indigenous and fishing communities) and environmental organizations (such as Greenpeace, Friends of the Earth), reinforce the need for accountability, citing ecological debt in Principles 4, 7, 8, and 9:

> [...] 4. Climate Justice asserts that governments are responsible for addressing climate change in a democratically accountable manner to their people and in accordance with the principle of common but differentiated responsibilities. [...] 7. Climate Justice requires the recognition of a principle of ecological debt owed by industrialized governments and transnational corporations to the rest of the world due to their appropriation of the planet's capacity to absorb greenhouse gases. 8. Affirming the principle of ecological debt, Climate Justice requires fossil fuel and extractive industries to be held accountable for all past and current impacts of the greenhouse gas production cycle and related local pollutants. 9. Affirming the principle of ecological debt, Climate Justice protects the rights of victims of climate change and associated injustices to receive full compensation, restoration, and reparation for lost lands, livelihoods, and other damages. (Bali Principles of Climate Justice, 2002, pp. 1–2)

Furthermore, the capabilities approach to justice, developed by Amartya Sen and Martha Nussbaum, emphasizes the freedoms necessary for individuals to lead functional and fulfilling lives. Schlosberg (2007) extends this approach to environmental justice, arguing that environmental protection is crucial for sustaining human capabilities. For instance, if literacy is a valued function, access to education is the necessary capability to achieve it. Similarly, in the environmental context, mobility and safety during extreme weather events are essential for maintaining a functional life amid climate change.

Recognition-Based Justice

Recognition is about acknowledging and valuing the identities, knowledge, and lived experiences of marginalized groups. In Latin America, the struggles of Afro-descendant and Indigenous peoples for recognition show how climate risks intersect with historical processes of racialization and dispossession. In the North, migrant communities or ethnic minorities often face similar misrecognition, where their vulnerabilities are rendered invisible in official statistics or adaptation plans. Recognition, therefore, becomes central to connecting justice to broader questions of equity and rights.

Intergenerational Justice

Finally, intergenerational justice raises the issue of long-term responsibilities. Climate change forces us to consider not only present vulnerabilities but also the rights and well-being of future generations. For coastal regions in Brazil, for example, rising sea levels pose existential risks for younger cohorts who may lose both their homes and cultural heritage. In Northern contexts, intergenerational justice often enters public debate through youth climate movements, such as *Fridays for Future*, which demand accountability from current governments and corporations.

Social Movements as Sites of Justice Claims

While this chapter emphasizes justice as the central analytical lens, social movements provide concrete illustrations of how distributive, procedural, and recognition justice are contested and redefined in practice. They are not treated here as the main object of analysis but as key arenas where the meaning of justice is articulated. In the Global South, grassroots organizations often mobilize around distributive justice, highlighting the unequal exposure to climate risks and environmental burdens. In Brazil, for example, favela residents and coastal communities have demanded adaptation strategies that address flooding, heat stress, and landslides, pointing to the concentration of risks in poor and racialized territories. Such claims show how climate justice cannot be separated from broader struggles against environmental racism and socio-spatial inequality.

At the same time, these movements engage in struggles for procedural justice, demanding meaningful participation in urban planning and adaptation initiatives. Experiences from participatory budgeting in Porto Alegre or climate action forums in Latin American cities illustrate attempts to democratize decision-making, though they often face structural barriers of exclusion.

Movements also reveal the centrality of recognition justice. Indigenous and Afro-descendant groups, for example, contest the erasure of their knowledge and cosmologies from dominant models of climate governance. By reframing climate change not only as a technical challenge but as a matter of cultural survival and historical reparation, they expand the scope of what "justice" entails.

In the Global North, climate justice movements have similarly foregrounded inequalities, though often in different forms. For instance, campaigns against energy poverty in European cities highlight distributive injustices, while movements for climate migrants' rights foreground recognition and procedural dimensions. Comparing North and South illustrates both the universality of justice claims and the specificity of their contexts: whereas in the South movements often address basic survival and historical inequities, in the North they may focus more on intergenerational justice or equitable transitions away from fossil fuels.

Taken together, these cases do not exhaust the diversity of social struggles but serve to demonstrate how climate justice is enacted, contested, and redefined across different geographies. They show that justice is not a fixed category but a lived and evolving practice, shaped by the interplay of social actors, institutions, and material environments.

Climate justice activism is deeply tied to global socioeconomic dynamics, as patterns of production and consumption drive climate change, disproportionately affecting vulnerable populations. Della Porta and Kriesi (1999) explore how globalization both constrains and enables social movements, with colonial forces and corporate influence shaping national responses while also facilitating transnational solidarity, knowledge-sharing, and visibility.

National contexts also remain crucial in determining a movement's trajectory. The political opportunity structure (POS) framework helps explain the success or failure of movements (Van Der Heijden, 1997). Della Porta (2013) argues that younger democracies, like Brazil, often fear political conflict due to historical experiences with authoritarianism, leading to more moderate movements. Other factors influencing movements include public bureaucracy, judicial autonomy, electoral instability, and government alliances. However, critics of POS highlight its overemphasis on political structures while neglecting cultural dynamics and the adaptability of modern movements beyond traditional party politics.

Other differences from North and South are seen in their radicality, characterized by disruptive tactics such as roadblocks, private property vandalism, and chaining oneself to strategic locations (Ostarek et al., 2024), which is more prevalent in the Global North and less common in Brazilian youth climate activism. These tactics exist across different types of movements, but social movements generally span a spectrum from moderate to radical approaches. It is also relevant to note that there's a proposed life cycle of movements in literature, as Christiansen (2011) and Della Porta and Diani (2006) discuss, in which they emerge, grow, institutionalize, and eventually dissolve (Fig. 2.1). This theoretical framework sets the stage for analyzing youth climate justice activism, particularly in Brazil, where both global and national factors shape movement strategies and political engagement.

Della Porta and Diani (2006) and Christiansen (2011) note that while certain organizations within a movement may follow distinct developmental

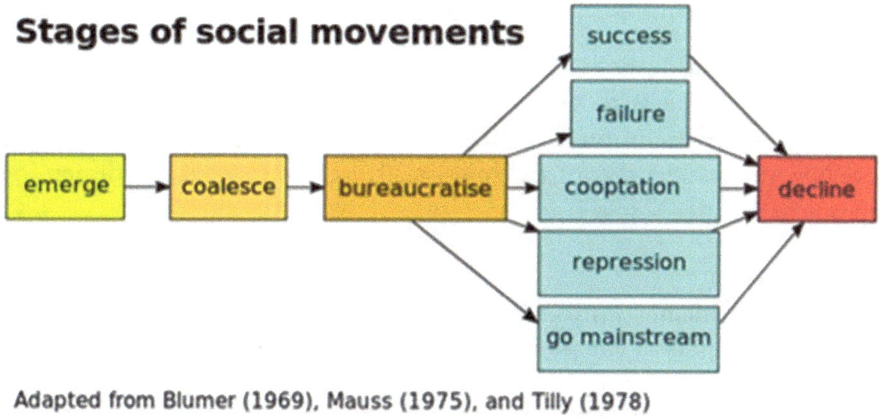

Fig. 2.1 A proposal for stages in the life cycle of social movements

stages, this does not always apply to the movement as a whole. "Generally, most movements do reach the stage of coalescence, since it is at that point that we begin to see behavior that we define as a social movement" (Christiansen, 2011, p. 23). However, with the rise of the internet, many movements stagnate at this stage and fail to institutionalize, whether by choice or circumstance. Regardless of the theoretical approach, social movements are inherently dynamic, influencing and being influenced by the world around them. This unpredictability is crucial in understanding the climate justice movement, where some organizations have institutionalized, others face repression, and new ones continue to emerge.

By examining justice as the core concept, this chapter highlights how climate risks and socio-ecological transformations are unevenly distributed across territories, classes, ethnicities, and generations. Rather than treating climate change as a purely environmental or technical challenge, a justice-oriented approach emphasizes questions of recognition, participation, redistribution, and intergenerational relations of responsibility (Schlosberg, 2007; Sultana, 2021).

To illustrate how these dimensions manifest in real-world contexts, Box 2.1 below invites students and other readers to reflect on concrete cases. The recent floods in Rio Grande do Sul, Brazil, in 2024, demonstrate how historical inequalities and institutional neglect shape who suffers most during climate disasters in the Global South. By contrast, the European heatwaves of 2022–2023 reveal that even high-income regions face challenges of recognition and preparedness, particularly for the elderly, migrants, and precarious workers.

These cases show that while the scale and intensity of vulnerabilities differ across North and South, the principles of climate justice remain central to understanding and responding to crises. They also reinforce the importance of comparing contexts to uncover both shared vulnerabilities and context-specific forms of exclusion. For students, this comparative lens is a way to move beyond abstract theory and to critically analyze how justice dimensions play out in practice across regions.

Box 2.1. Climate Justice in Practice: Comparing Extreme Events in the Global North and South

During consecutive summers, Europe experienced unprecedented heatwaves, with record-breaking temperatures exceeding 40 °C in several countries. The elderly, outdoor workers, and migrants without adequate housing were among the most affected. Cities such as Paris and Madrid struggled to adapt, revealing gaps in preparedness even in high-income contexts. Shortly afterwards, in May 2024, extreme floods devastated large areas of Rio Grande do Sul, Brazil, displacing hundreds of thousands of people. Vulnerable communities—particularly low-income, Black, and peripheral populations—were disproportionately affected, while wealthier neighborhoods recovered faster due to better infrastructure and insurance coverage.

These cases are merely a sampling of escalating extreme events attributed to climate change occurring around the world in recent years. What do these cases reveal about the different manifestations of vulnerability in the Global North and Global South? How does recognition justice (acknowledging marginalized groups) become relevant in climate crises in wealthier countries? Can comparing European heatwaves and Brazilian floods help us understand similarities and differences in climate justice challenges across contexts? To what extent did government and civil society responses address—or fail to address—climate justice in these cases? How do these events exemplify the need for both distributive and procedural justice in climate adaptation?

Climate Activism in Brazil

Social movements are diverse and dynamic, often encompassing both moderate and radical factions. Historical cases further illustrate these differences. In Brazil, the intersection of climate justice with racial justice is particularly significant, resembling labor and feminist movements in their resilience and structure. Acselrad (2010) highlights that international environmental organizations, such as Greenpeace Brazil, initially drove the country's environmental movement by introducing broad campaigns, funding, and external agendas. He distinguishes between "combative ecological movements," such as the rubber tappers of Acre, the Movement of People Affected by Dams (MAB), and the Landless Workers Movement (MST), which adopted less moderate repertoires. Alonso et al. (2007) note that since the 1970s, Brazil's environmental movement formed alliances with student groups, other social movements, cultural institutions, scientists, politicians, artists, and religious organizations. By the 1980s, environmental NGOs became more professionalized, focusing on lobbying and aligning with political parties.

Torres et al. (2020) compare climate mobilization in the Global North and South, noting that the sociological study of climate justice arrived late in Brazil. They highlight that the Fridays for Future movement, which organized the first Global Climate Strike in 2019, saw minimal participation in Brazil due to a lack of engagement with vulnerable communities. Besides, young climate activists face additional challenges in media representation. Studies show they are more frequently delegitimized compared to older activists, who are perceived as less radical (Neas et al., 2022; Rafaely & Barnes, 2020). Despite this, collaboration between different activist groups remains essential. Malm (2021) underscores the importance of radical factions in countering climate denialism, suggesting that strategic radicalism plays a crucial role in advancing climate justice.

Youth Climate Activism in the Netherlands

As outlined by Zantvoort (2023), the Netherlands is central to modern/colonial power structures, both as a colonizer and a key player in extractive industries like palm oil, soy, and cocoa. Yet, it is also highly vulnerable to rising sea levels due to climate change. One key difference in Dutch climate activism is the treatment of ethnicity and race, which is often seen as secondary, as highlighted by Van der Loeff (2022) and Rowe and Ormond (2023). The predominantly white population explains this, though the 13.2% immigrant population (CBS, n.d.) complicates the narrative. Climate justice also involves solidarity with others, particularly in light of the Netherlands' colonial history and its ongoing control of Caribbean territories like Aruba, Curaçao, and Saint Martin. Gender-wise, non-binary individuals consistently participate in climate activism in the Netherlands.

A commonality in activism between the Global North and South is repression, although violence against activists is more prevalent in Brazil. The Global Witness report (2024), though not exhaustive, shows that while repression exists in the Netherlands, it is less severe than in Brazil. Still, Dutch repression has increased, especially during protests that block major streets, also seen in the arrest of five climate activists in London in 2024 (Gayle, 2024). This aligns with Brazil, where large-scale street blockages are an often-used tactic by climate activists, such as Indigenous Peoples, cases in which police repression is widespread (Hanna et al., 2016).

Wahlström et al. (2019) highlight other differences in activism, such as Dutch youth activists' ability to discuss climate change with their parents and act together (reusing plastic bottles, donating to environmental groups, and protesting). In contrast, Talukdar (2022) describes immigrant youth activists in Australia struggling to discuss climate activism with unsupportive relatives.

Regarding their profile as protesters, Dutch youth activists mention feelings of anger, concern, and frustration as motivation, and they place the most trust in environmental groups and the least in the media, with a moderate level of trust in the police (Wahlström et al., 2019). Their discourse is shaped by intersectionality, colonialism, and historical reparations.

Particularities of Youth Climate Justice Activism Between North and South

Neas et al. (2022) argue that a sense of governmental failure to address climate issues leads to frustration and is a universal feeling among youth, driving them toward climate activism. In this sense, Schlosberg et al. (2024) observe that while some countries like Canada, France, and Australia offer state solutions, many activists in other regions seek solutions independent of the state. Della Porta and Portos (2023) show that high-income activists in Europe tend to trust governments more than middle and working-class activists, though the difference is small.

In the global South, other topics have more attention than the governments, such as the connection between youth climate justice organizations, like Fridays for Future Brazil, Engajamundo (2012) and Youth Climate Leaders (2018), with Indigenous peoples, landless communities, and dam-affected populations (Shirts, 2022). Or the intersection of gender inequality and climate change, examined by Sultana (2021) and Venegas et al. (2021) from a feminist Global South perspective. They argue that in these regions, the struggle for climate justice involves confronting colonialism, patriarchy, and capitalism.

Additionally, Barnwell and Wood (2022) point to the psychological consequences of climate injustices in the Global South, particularly the lack of mental health support. A literature review on South African research shows that while Global South countries contribute significantly to climate issues, the research focus remains heavily in the Global North, especially in the U.S.

Other social markers relevant to climate activists are ethnicity, race, migration background, physical ability, gender, and political affiliation. Louback (2020) addresses the role of ethnicity, race, and gender in Brazil, noting that Black women, who constitute 27% of the population, are invisible voices because there's no distinction for those social markers in climate policy. Indigenous women are made invisible as well, despite their leadership of environmental movements. For example, recently, in August 2025, the IV Indigenous Women's March in the capital Brasilia resulted in a declaration that calls for climate justice in public policies for indigenous populations, delivered to the National Congress (ANMIGA, 2025). Talukdar (2022) also highlights how ethnicity and religion influence climate activism in Australia, especially among South Asian youth, contrasting this with the experiences of white Australian activists.

Conclusion

This chapter has argued that climate change is fundamentally a question of justice, not only of technology or environmental management. By framing climate justice through distributive, procedural, recognition-based, and intergenerational dimensions, we have shown how climate risks and responses are unevenly distributed across groups, regions, and generations. The comparative cases from Brazil and the Netherlands illustrate both shared challenges and context-specific expressions of injustice: while in the Global South climate justice struggles are often about survival in the face of floods, heat, and displacement, in the Global North they frequently emphasize equitable transitions, intergenerational justice, and the politics of recognition for marginalized groups.

By connecting empirical cases of youth activism with theoretical frameworks, we highlighted how justice is not an abstract principle but a contested and lived practice. Social movements serve as arenas where different dimensions of justice are negotiated, revealing the centrality of ethnicity, race, class, gender, and colonial histories in shaping vulnerabilities and responses. At the same time, comparisons between North and South show that procedural failures, misrecognition, and unequal distributions of risks and resources are present in all contexts, albeit in different forms.

For students and other readers, the central lesson is that climate justice provides tools to critically evaluate not only who is most at risk, but also who participates in shaping climate futures and whose voices are acknowledged. Thinking through justice shifts climate change from a technical challenge to a deeply political and ethical one. It also highlights the need for solidarity across regions and generations, since just transformations require addressing historical inequities while building inclusive and democratic pathways for adaptation and mitigation.

Discussion Questions

Climate justice invites us to rethink climate change not only as a technical challenge but also as a deeply social and political issue. In your city or region, who is most exposed to climate risks such as heatwaves, floods, or droughts? Has there been a recent event (e.g., a heatwave, wildfire, or flood) where the impacts were unevenly distributed across different groups of people? Who gets to decide where adaptation projects—like flood defenses or green spaces—are located? Whose interests are prioritized? How might the answers to these questions differ if you compare countries in the Global North and Global South?

How do the different dimensions of justice explored here interact in climate adaptation and mitigation strategies?

Can participatory initiatives (such as budgeting processes or climate forums) truly ensure procedural justice, or do they risk reproducing existing exclusions?

Can you think of examples from your own city or country where risks or resources are unequally distributed? How might procedural or recognition failures reinforce distributive injustices?

In what ways do historical inequalities (colonialism, racism, uneven development) shape the meaning of climate justice in the Global South?

How do social movements in the Global South and Global North articulate different dimensions of justice (distributive, procedural, recognition)? What lessons can be drawn from comparing survival-oriented struggles in the South with intergenerational and transition-oriented struggles in the North?

References

Acselrad, H. (2010). Ambientalização das lutas sociais - o caso do movimento por justiça ambiental. *Estudos Avançados, 24*(68), 103–119. https://www.revistas.usp.br/eav/article/view/10469

Adger, W. N., Paavola, J., Huq, S., & Mace, M. J. (Eds.). (2006). *Fairness in adaptation to climate change.* MIT Press. https://doi.org/10.7551/mitpress/2957.001.0001

Agarwal, A., & Narain, S. (1991). *Global warming in an unequal world: A case of environmental colonialism.* Centre for Science and Environment. https://doi.org/10.1093/oso/9780199498734.003.0005

Alonso, A., Costa, V., & Maciel, D. (2007). Identidade e estratégia na formação do movimento ambientalista brasileiro. *Novos Estudos, 79*, 151–167. https://doi.org/10.1590/S0101-33002007000300008

Amorim-Maia, A. T., Anguelovski, I., Chu, E., & Connolly, J. (2022). Intersectional climate justice: A conceptual pathway for bridging adaptation planning, transformative action, and social equity. *Urban Climate, 41*, Article 101053. https://doi.org/10.1016/j.uclim.2021.101053

ANMIGA (Articulação Nacional das Mulheres Indígenas Guerreiras da Ancestralidade). (2025, August 7). *Carta dos Corpos-Territórios em Defesa da Vida.* http://anmiga.org/wp-content/uploads/2025/08/DOCUMENTO-FINAL-Carta-dos-Corpos-Territorios-da-IV-Marcha-das-Mulheres-Indigenas_.pdf

Arruda Filho, M. T., Torres, P. H. C., & Jacobi, P. R. (2024). A systematic review of the literature on climate justice: A comparison between the Global North and South. *Sustainability, 16*(22), 9888. https://doi.org/10.3390/su16229888

Barnwell, G., & Wood, N. (2022). Climate justice is central to addressing the climate emergency's psychological consequences in the Global South: A narrative review. *South African Journal of Psychology, 52*(4), 486–497. https://doi.org/10.1177/00812463211073384

Börner, S. (2023). Emotions matter: Empower-ing youth by integrating emotions of (chronic) disaster risk into strategies for disaster preparedness. *International Journal of Disaster Risk Reduction, 89*, Article 103636. https://doi.org/10.1016/j.ijdrr.2023.103636

CBS (Centraal Bureau voor de Statistiek). (n.d.). *Origin. Statistics Netherlands: How many residents of the Netherlands were born abroad?* Retrieved from https://www.cbs.nl/en-GB/visualisations/dashboard-population/origin

Christiansen, J. (2011). Four stages of social movements. In The Editors of Salem Press (Ed.), *Sociology reference guide: Theories of social movements*. Salem Press.

CorpWatch. (1999). *Greenhouse gangsters vs. climate justice*. Transnational Resource and Action Center (TRAC). https://www.corpwatch.org/sites/default/files/Greenhouse%20Gangsters.pdf

CorpWatch. (2002). *Bali principles of climate justice*. https://www.corpwatch.org/article/bali-principles-climate-justice

Della Porta, D. (2013). Repertoires of contention. In *The Wiley-Blackwell encyclopedia of social and political movements*. Wiley-Blackwell. https://doi.org/10.1002/9780470674871.wbespm178

Della Porta, D., & Diani, M. (2006). *Social movements: An introduction.* Blackwell Publishing.

Della Porta, D., & Kriesi, H. (1999). Social movements in a globalizing world: An introduction. In D. Della Porta, H. Kriesi, & D. Rucht (Eds.), *Social movements in a globalizing world* (pp. 3–22). Palgrave Macmillan. https://doi.org/10.1007/978-1-349-27319-5_1

Della Porta, D., & Portos, M. (2023). Rich kids of Europe? Social basis and strategic choices in the climate activism of Fridays for future. *Italian Political Science Review/rivista Italiana di Scienza Politica, 53*(1), 24–49. https://doi.org/10.1017/ipo.2021.54

Fraser, N. (2007). Re-framing justice in a globalizing world. In T. Lovell (Ed.), *(Mis)recognition, social inequality and social justice* (pp. 17–36). Routledge.

Gayle, D. (2024, July 18). Five just stop oil activists receive record sentences for planning to block M25. *The Guardian.* https://www.theguardian.com/environment/article/2024/jul/18/five-just-stop-oil-supporters-jailed-over-protest-that-blocked-m25

Global Witness. (2024). *Global Witness annual defenders report 2023/2024: Missing voices: The violent erasure of land and environmental defenders*. https://www.globalwitness.org/en/campaigns/environmental-activists/missing-voices/

Hanna, P., Langdon, E. J., & Vanclay, F. (2016). Indigenous rights, performativity and protest. *Land Use Policy, 50*, 490–506. https://doi.org/10.1016/j.landusepol.2015.06.034

Juhola, S., Heikkinen, M., Pietilä, T., Groundstroem, F., & Käyhkö, J. (2022). Connecting climate justice and adaptation planning: An adaptation justice index. *Environmental Science & Policy, 136*, 609–619. https://doi.org/10.1016/j.envsci.2022.07.024

Kashwan, P. (2021). Climate justice in the Global North. *Case Studies in the Environment, 5*(1). https://doi.org/10.1525/cse.2021.1125003

Kuehn, R. R. (2000). A taxonomy of environmental justice. *Environmental Law Reporter, 30*, 10681–10703. https://ssrn.com/abstract=628088

Louback, A. C. (2020, July 31). O paradoxo da justiça climática no Brasil: O que é e para quem? *Le Monde Diplomatique Brasil*. https://diplomatique.org.br/o-paradoxo-da-justica-climatica-no-brasil-o-que-e-e-para-quem/

Malm, A. (2021). *How to blow up a pipeline: Learning to fight in a world on fire*. Verso.

Neas, S., Ward, A., & Bowman, B. (2022). Young people's climate activism: A review of the literature. *Frontiers in Political Science, 4*. https://doi.org/10.3389/fpos.2022.940876

Ojala, M., et al. (2021). Anxiety, worry, and grief in a time of environmental and climate crisis: A narrative review. *Annual Review of Environment and Resources, 46*(1), 35–58. https://doi.org/10.1146/annurev-environ-012220-022716

Ostarek, M., Simpson, B., Rogers, C., & Ozden, J. (2024). Radical climate protests linked to increases in public support for moderate organizations. *Nature Sustainability, 7*(12), 1626–1632. https://doi.org/10.1038/s41893-024-01444-1

Rafaely, D., & Barnes, B. (2020). African climate activism, media, and the denial of racism: The tacit silencing of Vanessa Nakate. *Community Psychology in Global Perspective, 6*(2/2), 71–86.

Rowe, T., & Ormond, M. (2023). Holding space for climate justice? Urgency and 'regenerative cultures' in extinction rebellion Netherlands. *Geoforum, 146*, Article 103868. https://doi.org/10.1016/j.geoforum.2023.103868

Schlosberg, D. (2007). *Defining environmental justice: Theories, movements, and nature*. Oxford University Press. https://doi.org/10.1093/acprof:oso/9780199286294.001.0001

Schlosberg, D., & Collins, L. B. (2014). *From environmental to climate justice: Climate change and the discourse of environmental justice*. Wires Climate Change. https://doi.org/10.1002/wcc.275

Schlosberg, D., Rickards, L., Pearse, R., Della Bosca, H., & Moraes, O. (2024). Critical environmental justice in contemporary scholarship and movements: Consensus and plurality of the discourse. *Environmental Politics, 34*(3), 399–420. https://doi.org/10.1080/09644016.2024.2362573

Shirts, M. (2022). *Emergência climática: O aquecimento global, o ativismo jovem e a luta por um mundo melhor* (1a ed.). Claro Enigma.

Sultana, F. (2021). Critical climate justice. *The Geographical Journal, 188*(1), 118–124. https://doi.org/10.1111/geoj.12417

Talukdar, R. (2022). *Why North-South intersectionality matters for climate justice: Perspectives of South Asian Australian youth climate activists* [Report]. Sapna South Asian Climate Solidarity.

Torres, P. H. C., Jacobi, P. R., & Leonel, A. L. (2020). Nem leigos nem peritos: O semeador e as mudanças climáticas no Brasil. *Política & Sociedade, 19*(44), 17–38. https://doi.org/10.5007/2175-7984.2020v19n44p17

Torres, P. H. C., Urbinatti, A. M., Gomes, C., Schmidt, L., Leonel, A. L., Momm, S., & Jacobi, P. R. (2021). Justiça climática e as estratégias de adaptação às mudanças climáticas no Brasil e em Portugal. *Estudos Avançados, 35*(102), 159–176. https://doi.org/10.1590/s0103-4014.2021.35102.010

Van der Heijden, H.-A. (1997). Political opportunity structure and the institutionalisation of the environmental movement. *Environmental Politics, 6*(4), 25–50. https://doi.org/10.1080/09644019708414357

Van der Loeff, A. S. (2022). *White inside: A critical examination of how structural whiteness in the Dutch climate movement obstructs climate justice* (Master's thesis, Lund University, Lund, Sweden). http://lup.lub.lu.se/student-papers/record/9097259

Venegas, M. M., Lang, M., & Jurado, G. R. (2021). *Climate justice from the perspective of Latin American and other Southern feminisms.* Rosa-Luxemburg-Stiftung, Regional Office Andean Region.

Wahlström, M., Kocyba, P., De Vydt, M., & de Moor, J. (Eds.). (2019). *Protest for a future: Composition, mobilization, and motives of the participants in Fridays For Future climate protests on 15 March, 2019 in 13 European cities.* https://doi.org/10.17605/OSF.IO/XCNZH

Zantvoort, F. (2023). Movement pedagogies in pandemic times: Extinction Rebellion Netherlands and (un)learning from the margins. *Globalizations, 20*(2), 1–14. https://doi.org/10.1080/14747731.2021.2009319

3

Inequality

Jennifer E. Givens and Mufti Nadimul Quamar Ahmed

Introduction

Climate change and related environmental problems threaten life as we know it on our planet. What is to be done? This chapter explores what environmental sociology has to say about the interrelated concepts of inequality and inequity and how this research relates to a just transformation. Environmental sociologists focus on inequality as a key component of the climate crisis. Examining this sometimes-overlooked concept helps us accurately understand the causes and consequences of social-environmental problems. For example, at an individual level, wealthier people, and at an international level, wealthier nations of the Global North, have done comparatively more to cause climate change, while less wealthy people and less wealthy nations of the Global South often suffer the worst consequences. Furthermore, research shows inequality itself drives climate change and obstructs effective climate action. Specific literatures in environmental justice and climate justice focus on societal inequities and call for more equitable ways forward. The relationship between inequality and multiscale drivers and impacts is important to understand if we want to effectively and equitably address climate change.

J. E. Givens (✉) · M. N. Q. Ahmed
Utah State University, Logan, UT, USA
e-mail: jennifer.givens@usu.edu

M. N. Q. Ahmed
e-mail: mufti.ahmed@usu.edu

D. J. Davidson et al. (eds.), *Environment and Society*,
https://doi.org/10.1007/978-3-032-27059-7_3

This chapter reviews key questions, areas of research, and theoretical perspectives related to inequality and the environment and discusses their relevance to the concept of just transformation.

Background

Sociology is the study of society, and the discipline focuses on various types of inequality across societies. Scholars examine the causes and consequences of inequalities between different groups, in different contexts, and at multiple scales from the local to the global. Scholars study inequality related to class, gender, age, sexuality, queer identities, disability status, Indigeneity, race/ethnicity, nationality/citizenship, migration status, Global North/Global South context, other social factors and social structures, and intersections between these factors (Dietz et al., 2020). While sociology is the study of society, environmental sociology explores environment and society relationships, in which inequality is just as central. Environmental sociologists examine individual and community contexts and social systems, especially "the ways that social and structural inequality play foundational roles in shaping environmental harms, environmental injustices, and potential solutions" (Caniglia et al., 2021, p. 4). Research shows that responses to address environmental issues and climate change should address such inequalities in order to be effective and to remedy, rather than further exacerbate, societal inequity (Dietz et al., 2020). Thus, attending to (in)equality and (in)equity is vitally important, and research in environmental sociology directly contributes to this effort.

What is inequality, and why do we care about it? Inequality describes situations in which things are unequal, whereas equality is when things are equal. It may be helpful to think about three types of hypothetical societal equality. First, everyone could be considered equal if they have the same status as a unique individual, have the same rights, and face the same rules of the system. Second, everyone could be considered equal if they have the same opportunities within the system. Third, everyone could be considered equal if they have the same outcomes. Whether or not each of these *is* true can be supported or refuted with empirical evidence. Whether or not each of these *should* be true can also be supported, or not, with empirical evidence, but these are normative questions, or value-based judgments. Thus, *if* these should be societal goals, and if so, *how* they could be achieved, are often debated. This leads to the related term, equity, which means having the quality of being fair and just. These are also normative concepts and thus open to debate. Empirical evidence from environmental sociologists helps to inform debates about inequality and equity in society, including how they relate to climate change and possibilities for a just transformation.

Drivers

What causes environmental problems such as climate change? A rich body of literature in environmental sociology examines the anthropogenic drivers of climate change, or in other words, the "social systems and human actions causing emissions, and societal factors shaping and conditioning those actions" (Jorgenson et al., 2024). This literature finds economic growth is a major driver (Jorgenson & Clark, 2012; Kallis, 2011; Kallis et al., 2018; Schor & Jorgenson, 2019). Since economic growth and development have varied historically and globally, this is a key source of the inequality in climate change causes and impacts.

Research also finds that economic inequality itself is a driver of climate change, including income and wealth inequality (Roberts, 2001; for reviews see Givens et al., 2023; Jorgenson et al., 2019). Research finds income inequality is associated with higher production-based and consumption-based carbon emissions among high-income nations (Jorgenson et al., 2017; Knight et al., 2017) and in lower-income nations (Hubacek et al., 2017). Similarly, but at the sub-national scale, in the U.S. income concentration at the top is associated with higher carbon dioxide emissions, although findings vary for general inequality (Adua, 2022; Jorgenson et al., 2017). Research also finds that addressing inequality at different scales or in different contexts may have differing effects on climate change, increasing emissions and other environmental problems. For example, reducing inequality by increasing the incomes of households with the lowest incomes could potentially increase emissions by increasing consumption (see Jorgenson et al., 2024). In other words, addressing equity issues could potentially exacerbate environmental issues, just as addressing environmental issues without attention to equity could exacerbate inequality. Thus, it is important to consider ways to avert this, such as policies that incorporate contraction and convergence approaches that involve reductions at the very top of income scales to address unequal overconsumption of resources. Although addressing inequality at different scales or in different contexts may have differing effects on climate change and other environmental problems, overall, much research finds an association between various measures of societal inequality and environmental impact. Increasingly, research explores links between emissions and other types of inequality such as gender inequality (McGee et al., 2020), working hours (Fitzgerald, 2022), or economic sector-specific contexts (Huang, 2024a, 2024b).

Environmental sociological research that employs political-economy perspectives, including some of the studies mentioned above, illuminates how income inequality may increase emissions and other environmental damage, drawing attention to how wealthy entities, from individuals to nation-states, reap disproportionate benefits from the environmentally damaging global political economic system. This economic and power inequality encourages maintenance of the status quo rather than actions to address climate change or improve societal inequity (for reviews see Givens et al., 2023; Jorgenson et al., 2024). One mechanism by which this occurs is that advantaged individuals and entities have more power to exert influence in the political sphere such as by lobbying against environmental protections (Brulle, 2021; Farrell, 2016). Stoddard et al. (2021) delineate how inequality is a driver that acts as a barrier to effective climate action because it separates the more vulnerable from the more powerful, it erodes the social trust between these groups, it reinforces elite preferences for the status quo as opposed to changes to address environmental issues or equity, and it allows elites to protect these interests. More systemically, some scholars see capitalist dynamics driving the whole global economic system toward more extreme inequality and wealth concentration (e.g., Piketty, 2014a, 2014b). Other scholars examining the capitalist global political economic system identify this treadmill of capitalist accumulation as the driver of the ecological crisis (Foster, 1999; Foster et al., 2010, 2011). Those studying patterns of consumption provide further insight into the nature of the association between inequality and the environment, with inequality driving status consumption, where more is always needed because status is relative, and this is exacerbated by marketing and cultural norms; this leads to disproportionate impacts of the rich, while at the same time there is globally uneven underconsumption with people unable to meet basic needs (Barros & Wilk, 2021; Rieger & Schor, 2021).

In addition to the ideas discussed so far, several theoretical perspectives and associated bodies of research, discussed next, represent rich examinations of the relationship between inequality and environmental issues such as climate change. Each of these theoretical perspectives and the empirical research to support them are vital in providing a more accurate understanding of inequality as it relates to climate change and the environment. These perspectives help counter dominant paradigms that do not adequately address equity or the environment, and provide varying ideas for possible alternative ways forward.

Treadmill of production theory is a foundational theory in environmental sociology that attends to the workings of the capitalist system and its impact on inequality and the environment (Schnaiberg, 1980). According to this theory, capitalism is premised on economic growth, and economic growth inevitably causes environmental deterioration due to the ever-increasing demand for resources and the creation of waste (see Gould et al., 2004; Jorgenson, 2016a). Those in charge of production are driven by the inherent logic of the system to increase their capital investment and replace labor with technology. This creates profits for those in charge of production but has detrimental effects on workers, leading to job displacement and increasing inequality (Gould et al., 2004). The theory argues for focusing on production more than consumption and posits that adopting a more democratic approach to the ownership and control of production could better solve social and ecological problems (Gould et al., 2004). Empirical work has tested this theory (for a review, see Jorgenson, 2016a).

Work in what is known as the *IPAT or STIRPAT* tradition has long focused on how inequality shapes environmental impact. This perspective, with roots in environmental sociology, structural human ecology, and ecological economics, conceptualizes environmental impact as a product of the interplay between population, affluence, and technology (Dietz & Jorgenson, 2013; Rosa & Dietz, 2012; York et al., 2003). Thus, inequality is considered to be inherently related to varying levels of environmental impacts across society (for a review, see Dietz & York, 2021).

Research also explores additional inequalities related to other drivers, including urbanization, corporations, and militarization (for an interdisciplinary review see Jorgenson et al., 2019; see also Jorgenson et al., 2024). For example, there is inequality in corporate emissions: a comparatively small number of corporations contribute a disproportionate amount to global greenhouse gas emissions (Grant et al., 2020). Thus, addressing disproportionality and the inequality across these super polluters could be a sustainability strategy. Other factors such as elite control of corporations, corporate norms, and corporate position in the unequal global economic system serve to reproduce the status quo of the unequal production of environmental harms (Downey, 2015; Ladegaard & Rieger, 2024; Rieger, 2024). Inequalities in the size and capital intensiveness of nations' militaries are also important for understanding carbon emissions, especially because militaries produce high levels of emissions even when not actively engaged in conflict (Hooks et al., 2021; Jorgenson et al., 2023). Addressing unequal global militarization as a driver of climate change and environmental impact may be difficult, but it is important to consider this very unequal institution in plans to address climate change and related issues.

Treadmill of destruction theory builds on treadmill of production theory but puts the focus on militarization. Hooks and Smith (2004) introduced this theory to help explain environmental inequalities experienced by Native Americans in the United States as a consequence of military operations. This idea adds military dynamics, with their own unique growth logic, to the focus on production treadmills. This perspective has been applied at various scales. For example, in related work Jorgenson et al. (2023) examine two areas of inequality in global militaries. These are military expenditures per soldier, which indicates how capital-intensive a nation's military is, and military participation rate, which is the proportion of a country's workforce that serves in the military. The authors find both are positively associated with carbon emissions, highlighting the importance of unequal military power globally. Smith et al. (2020) apply the two treadmill perspectives at a more local scale to investigate gold mining and coca/cocaine production in Colombia and Peru. In Colombia, the authors observe that a protracted internal conflict resulted in synergistic treadmills of production and destruction, as guerrilla and paramilitary organizations used the industries to finance their operations, leading to environmental destruction as part of the conflict. While armed groups have demobilized since the 2016 ceasefire, resulting in a decrease in the treadmill of *de*struction, a violent *pro*duction treadmill has intensified, focused on gold mining and other industries, with a particular emphasis on land grabbing and resource exploitation by influential economic actors who are exploiting the power vacuum in former conflict zones. In Peru, gold mining is characterized by a clear and highly destructive treadmill of production controlled in part by criminal organizations, with both legal and illegal mining operations resulting in significant environmental harm (Smith et al., 2020). These studies demonstrate how some of these theories are complementary in drawing our attention to various forms of inequality, such as inequalities in development or militarization, and they demonstrate the importance of also considering how inequalities interact with various forms of power.

Similar to Treadmill of Production Theory, *Metabolic rift theory* emphasizes production, but in contrast to the Treadmill of Production, draws more attention to the spatial or geographic as well as environmental and other forms of inequality driven by the global capitalist system of production. It examines how the capitalist political economy drives ecological imperialism, disrupting natural processes and cycles, driven by and perpetuating structures of spatial and political economic inequality. The theory is attributed to Marx and his analysis of soil fertility and agriculture under capitalism (see Clark & Foster, 2009; Foster, 1999). Clark and Foster (2009) analyze the rift on a global scale, examining the large-scale removal of guano and nitrates from Peru and Chile to satisfy European agricultural demands. Degradation of the environment, exploitation of workers, economic dependency, and war were some of the consequences of this trade for Peru and Chile. Focusing on counter processes with the potential to address inequalities and metabolic rifts, Wittman (2009) examines the global food sovereignty movement at the national and sub-national scale, represented by organizations like La Vía Campesina and its Brazilian affiliate the Landless Workers Movement, suggesting agrarian citizenship as a way to mend the rift, address inequality, and improve the lives of rural residents. Likewise, McClintock (2010) examines the idea of urban agriculture from the metabolic rift perspective and contends it too can help heal three interrelated rifts: ecological, social, and individual.

Ecologically unequal exchange theory and the concept of environmental load displacement examine inequalities in the global system in terms of power and economic relationships, and the specific contribution of this perspective is that it shows how unequal global trade relationships drive environmental harm and displace it across space and time. In other words, these perspectives focus on how trade, foreign investment, and other aspects of the unequal global political economic system are mechanisms that structure unequal environmental benefits and burdens globally. *Ecologically unequal exchange* posits that the market pricing of primary products exported by economically disadvantaged nations frequently fails to account for the environmental and social costs linked to their extraction, processing, and transportation. This leads to an inequitable trading relationship in which Global North nations benefit from advantaged positions in trade, supporting elevated consumption and distancing from the impact of this consumption, while less affluent nations endure the ecological and economic detriments (see Bunker, 1984; Givens & Huang, 2021; Givens et al., 2019; Hornborg, 2009; Jorgenson, 2006, 2009, 2016a, 2016b; Roberts & Parks, 2009). In this global relationship characterized by inequality, Global South nations are treated as a source of resources and a destination for waste within the global economic system (Givens et al., 2019). These mechanisms generate *environmental load displacement*, which occurs over time and displaces the burdens of Global North production and consumption geographically and temporally (Muradian et al., 2002). From an equity and justice perspective, this system also generates an ecological debt. In order to address this debt, some argue developed countries should compensate less developed nations for the harm their resource-intensive production and consumption cause to the environment (Roberts & Parks, 2009; Warlenius, 2018). Foundational work in these perspectives (Hornborg, 1998; Jorgenson, 2006, 2009; Muradian et al., 2002) has fostered a large body of research that tests hypotheses based on the theory of ecologically unequal exchange and frequently finds support for the theory across multiple environmental issues, including deforestation, greenhouse gas emissions, biodiversity loss, water pollution and scarcity, and ecological footprint (see Givens & Huang, 2021; Givens et al., 2019).

Finally, the *carbon intensity of well-being (CIWB) measure* is an example of a concept from environmental sociology that captures inequalities at multiple scales. CIWB, often operationalized as a ratio of carbon emissions to life expectancy, represents how carbon intensely countries or other entities create well-being for residents (Dietz & Jorgenson, 2014; Jorgenson & Dietz, 2015; Jorgenson & Givens, 2015). This indicator specifically captures the unequal benefits and burdens of development and climate change. Looking at how country-level income inequality affects CIWB, Jorgenson (2015) finds that CIWB began increasing in the 1990s, and the effect continues to increase in magnitude. The takeaway from this is that reducing income inequality could have beneficial effects for environmental conditions, climate change mitigation, and human well-being. In a U.S.-based analysis, Jorgenson et al. (2018) note that sustainability requires reducing the CIWB. They find, furthermore, that higher rates of inequality and poverty are associated with higher CIWB, again indicating that reducing inequality would benefit both the environment and human well-being. Other research looks at various asymmetric effects on CIWB including the effects of economic growth and fossil fuel dependence at global and US scales (Greiner & McGee, 2020; Thombs et al., 2022) and the effects of other inequalities such as gender and racial inequality (Briscoe et al., 2021; Ergas et al., 2021) and urbanization (Givens, 2015; McGee et al., 2017; Wang et al., 2022) across scales. In a paper combining CIWB and ecologically unequal exchange theory, Givens (2018) investigates the impact of global trade integration and trade linkages on the CIWB of 81 nations between 1990 and 2011 and finds a higher CIWB in less developed nations is associated with greater exports to high-income countries. In another paper exploring inequalities in the level of national integration into the world system, Givens (2017, p. 403) explores the impact of world society and polity integration on CIWB, finding mixed results based on a country's Global North versus Global South status, which highlights "the complexities of sustainable development in an unequal global system." The CIWB is a key concept for analyzing inequalities in greenhouse gas emissions and well-being.

Unequal Impacts

The above section focuses on drivers of environmental harm including structural characteristics of our unequal political economic system and how this relates to the concept of inequality in environmental sociology. In this next section, we explore two bodies of literature that focus especially on inequalities in impacts and calls for more equitable ways forward. These are the literatures on environmental justice and climate justice. While the concept of justice is given close attention in another chapter, here we highlight the intersections between justice and (in)equality.

Environmental justice is a concept and a broad area of research with some of its beginnings in sociology. Topics include studies of environmental justice movements and equitable or inequitable processes for justice, including addressing historical and structural relationships that shape injustices (for reviews see Agyeman et al., 2016; Mohai et al., 2009). The work increasingly incorporates scholarship on Indigenous epistemologies and includes justice for more than human species (Liévanos et al., 2021; Schlosberg, 2013; Whyte, 2020). Others take on the role of the state as complicit in environmental injustice (Pellow, 2017; Purucker, 2021). The environmental justice literature often focuses on the local level, although the work has also globalized. Work on the "environmentalism of the poor" is foundational to environmental sociology and environmental justice (Anguelovski & Alier, 2014; Martinez-Alier et al., 2016), and the expanding body of work on the EJAtlas, which documents environmental and climate justice conflicts around the world, is another example of the globalization of the literature on environmental justice (EJAtlas, 2024; Martinez-Alier et al., 2016; Temper et al., 2018), as is work that frames ecologically unequal exchange as a theory of global injustice (Givens et al., 2019).

Because the concept of justice is focused on the *rectification* of inequities and inequalities, this work emerges from, and continues to be closely related to, research on how social systems structure and maintain environmental inequality. While environmental justice research began by documenting racial disparities in exposures to toxic waste (Bullard, 1983, 1990), the research has deepened, broadened, and diversified. Environmental justice research today is grounded in studies that examine how environmental inequalities follow existing patterns of inequality in society shaped by class, gender, age, sexuality, queer identities, disability status, Indigeneity, race/ethnicity, nationality/

citizenship, migration status, Global North/Global South context, and other social factors. Recent work increasingly explores how intersectionality shapes justice (Dietz et al., 2020; Ergas et al., 2021; Malin & Ryder, 2018; Ryder, 2018), another concept that is the focus of a different chapter in this volume.

The study of climate justice similarly emerges from "the concern that the causes and consequences of climate change and the impacts of efforts to reduce the magnitude of climate change and adapt to it, are inequitably distributed" (Dietz et al., 2020, p. 144). Climate justice builds on foundational insights about inequality from environmental justice scholarship and activism and is unique in its greater emphasis on unequal systemic and structural political economic perspectives; and while environmental justice movements tend to grow from local, community focused cases at local scales, climate justice movements span scales from the local to the global, from local decarbonization movements to the global climate youth movement (Mohai et al., 2009). Early foundational research in sociology on climate inequality and justice documented global political economic inequality in contributions to global climate change versus exposure to its impacts, the example we began with in this chapter (Roberts & Parks, 2006, 2009). Scholarship in climate justice now spans communities around the world facing the impacts of climate change including physical displacement, threats to livelihoods, and adaptation strategies (Ahmed et al., 2024; Falzon & Sen, 2024), to uneven flows of resources, global climate negotiations, global climate finance, climate change mitigation, and obstruction to climate action (Falzon et al., 2021). Mechanisms such as ecologically unequal exchange may act as a catalyst for climate justice movements as those impacted experience the harmful environmental effects and increasingly become aware of the global inequality being enacted (Almeida, 2023; Fisher, 2024; Jorgenson & Givens, 2014; Jorgenson et al., 2024). Studies in this area also include attention to inequalities and injustices inherent in some strategies to address climate change (Markkanen & Anger-Kraavi, 2019), and others make the case for the vital need to consider equity in climate policy (Klinsky et al., 2017). As with environmental justice, climate justice scholars increasingly incorporate human-animal relationships, non-human species, and multispecies justice (Dietz et al., 2020; Guenther, 2024; Taylor et al., 2018). Additionally, as the connections between climate change and other environmental issues grow, environmental and climate justice increasingly overlap (Maung & Pellow, 2021).

Reform Versus Transform and Possibilities for Just Transformation

Because inequality is a driver of environmental harm including climate change and it is a factor blocking communication and collaboration toward effective and equitable climate action, addressing inequality must be part of the response to these issues. Responses that do not address inequality risk not being effective, exacerbating inequality, and being unjust. All of the theoretical perspectives and empirical research described above can help to counter more dominant paradigms that do not adequately address the relationships between environmental issues and inequalities.

Ecological modernization theory is a final theory to discuss. For a time, it was a key theory in environmental sociology and it remains a prominent paradigm in environmental policy discussions, but here, we draw attention to how it neglects to adequately theorize inequality, the key focus of this chapter, and by extension the implications of its popularity for Just Transformation. Ecological modernization theory suggests that ongoing economic progress can help alleviate environmental challenges as industrialized societies begin to emphasize sustainability through technological advancement and policy improvements (see Fisher & Freudenburg, 2001; Jorgenson, 2016a; Spaargaren & Mol, 1992; York et al., 2010). This theoretical perspective arose in the late twentieth century as a response to the growing need to comprehend how contemporary society's institutions could be transformed to attain environmental sustainability. According to Spaargaren and Mol (1992), the theory emphasizes the industrial nature of modern society rather than its capitalist nature, and the theory is a theory of social change. Fisher and Freudenburg (2001) explain that the theory assumes economic entities will take a prominent position in promoting ecological transformations, and it envisions political players banding together to make environmental preservation a realistic political possibility. However, since the emergence of ecological modernization theory it has been tested and critiqued. Carolan (2004) criticized the sole emphasis on production and argued that to alleviate environmental issues consumption must also be problematized, yet making "green" decisions based on the premise of a "knowledgeable consumer" is difficult due to the complexity of global commodities networks, which disguise social and environmental inequalities and consequences of purchase choices (Carolan, 2004). York and Rosa (2003) argue that the theory needs to show actual impacts, frequently enough and at a large enough scale to matter, without creating environmental load displacements to other industries, firms, entities, or locations, and without a Jevons Paradox in which more efficiency just means more consumption, which in turn means more environmental problems. While this perspective may have been a hopeful attempt to document a green transition, ecological modernization theory ignores inequalities and injustices in the system and is more frequently associated with uncritical neoliberal beliefs in the benefits of free trade and possibilities for green growth or a green economy. This belief serves to maintain the inequitable status quo by benefiting those holding concentrated wealth and power in the current system. Such perspectives neglect the significant inequalities that other perspectives in environmental sociology so effectively reveal.

Despite the problems with the vision of social change proposed by ecological modernization theory, the key question remains: *how can* our society address and overcome the ecological crisis it has created? If not through economic and technological development, how might it occur? Environmental sociology's contribution is in part to highlight the importance of addressing inequality in addressing this question. Although some policies attempt to mitigate climate change or lead to a more sustainable society, they may not address and may further exacerbate inequalities. A *just transformation* would "ensure equitable processes and outcomes in creating a more just, sustainable, and less-carbon intensive society" (Kojola & Agyeman, 2021, p. 115). A just transformation would address social justice *and* the myriad environmental issues we face. Research in environmental sociology contributes by providing an empirically based understanding of the multiscalar and various forms of inequality inherent in our system and reviewed in this article, including the structural and systemic barriers that will need to be addressed to achieve a just transformation. Furthermore, the pursuit of a just transformation would bring in further sociological insights on additional processes of colonialism, racism, and patriarchy (see Kojola & Agyeman, 2021).

Finally, there is a growing body of research on just transformation itself from environmental sociologists. The concept of 'just transformation' is rooted in the labor and environmental justice movement in the United States and envisions moving to a low carbon economy, climate resilience, and a system that is equitable and inclusive (Ciplet, 2022). A just transformation prioritizes the demands of workers and underprivileged populations in sustainability initiatives (Ciplet & Harrison, 2020). In the 1970s, those who advocated for 'just transition' mostly sought to improve the working conditions of chemical and fossil fuel industry workers, but increasingly the focus has shifted from workers to the "frontline" populations that are bearing the brunt of environmental damage (Ciplet, 2022). In research on the possibility for a just transition, Ciplet and Harrison (2020) identify three fundamental tensions inherent in just transition initiatives, drawing on interviews with environmental justice, Indigenous, and just transition movement leaders. First, sustainability-inclusivity tension arises when balancing the speed at which we need to adopt effective environmental policies while simultaneously guaranteeing inclusive decision-making processes that consider all stakeholders' viewpoints. Second, the sustainability-recognition tension denotes the difficulty of balancing uniform sustainable performance metrics with the acknowledgment and protection of different human rights and value systems, especially those of historically oppressed groups. Third, the sustainability-equity tension results from the possible contradiction between reaching sustainability targets and equitably distributing the advantages and negative effects related to those initiatives. For each of these conflicts, the authors provide four possible outcomes ranging from scenarios in which justice is neglected in favor of sustainability to ones in which both sets of goals are satisfied, indicating a just transition. In other work, Ciplet (2022, p. 315) utilizes sociological theories about power and social change, including ones from Antonio Gramsci and Karl Polanyi, to analyze the potential of calls for just transformations to be truly transformative "in shifting the political economic structures that cause, sustain, and deepen injustices." He develops a typology ranging from those that maintain the status quo to those that could be transformative, but notes that transformation potential depends on building diverse coalitions that are difficult to form and sustain, especially in the face of structural obstacles and opposition from dominant hegemonic coalitions that preserve the status quo. In other research from environmental sociologists related to just transformations, Pellow (2017) and Harrison (2019) call into question the ability of the state under the current capitalist political economic system to be an ally in environmental justice efforts and rather see the state as complicit in environmental injustice or ineffective at moving toward justice. Research in environmental sociology should continue to examine barriers to and possibilities for just transformation moving forward.

Conclusion

The concept of inequality—how it is understood, measured, and addressed (or not)—has deep implications for collective efforts to confront climatological and environmental crises through just transformation. There are inequalities in climate change drivers and consequences at multiple scales and across space and time that must be accounted for. Inequality itself is a driver of climate change, suggesting that failures to address inequalities can compromise Just Transformation ambitions. Inequality also blocks effective action to mitigate and adapt to climate change. Environmental sociological research, because of its disciplinary focus on inequality and equity, and its tradition of examining social structure, is well suited to identifying alternatives to our modern inequitable and unsustainable world (Klinenberg et al., 2020) and offering useful perspectives on the possibility to move toward a just transformation. The discipline has the necessary theoretical tools and methods to question and examine the potential for system change (Dietz et al., 2020). Environmental sociology provides theoretical perspectives and empirical evidence for drivers, structural barriers, and inequalities that will be necessary to address in order to work toward just transformation.

Reflection and Discussion Points

Is inequality acceptable? How much? Are some types of inequality more acceptable than others? What types of inequality constitute inequity? How might attempts to address inequity, and by extension pursuing a just transformation, negatively impact the environment?

What does just transformation look like to you? What are some barriers to your vision of a just transformation? How can the sociological theories described above help you think through both barriers and opportunities? Apply one of the theoretical perspectives to your vision of a just transformation. How does this perspective inform your vision?

Think about various proposed solutions you have heard for addressing climate change, such as a carbon tax. Do these solutions address inequality? If so, how? Do they move us toward a just transformation? If not, how could addressing inequality be incorporated to better make progress toward a just transformation?

Think about the topics, theoretical perspectives, and approaches covered in this chapter. What further research questions do you have? Design a research project utilizing content from this chapter. How would your research advance our understanding of a just transformation?

If you were working on policy to address climate change, what are three key takeaways from this chapter, and how might you begin to implement them in policy change? Propose a policy to address climate change, and then critically examine its ability to move toward a just transformation. How could you refine the policy to better address inequality and move toward a just transformation?

What are some things you can do at the individual level to address inequality and move toward a just transformation?

Brainstorm some visions of change that are more systemic, or at the system level, such as the political economic system, that could move us toward just transformation. What are the barriers to this sort of change? Where are the opportunities?

References

Adua, L. (2022). Super polluters and carbon emissions: Spotlighting how higher-income and wealthier households disproportionately despoil our atmospheric commons. *Energy Policy, 162*, Article 112768. https://doi.org/10.1016/j.enpol.2021.112768

Agyeman, J., Schlosberg, D., Craven, L., & Matthews, C. (2016). Trends and directions in environ-mental justice: From inequity to everyday life, community, and just sustainabilities. *Annual Review of Environment and Resources, 41*, 321–340. https://doi.org/10.1146/annurev-environ-110615-090052

Ahmed, M. N. Q., Givens, J. E., & Arredondo, A. (2024). The links between climate change and migration: A review of South Asian experiences. *SN Social Sciences, 4*(3), 64. https://doi.org/10.1007/s43545-024-00864-2

Almeida, P. (2023). Protest waves in Latin America: Facilitating conditions and outcomes. In F. M. Rossi (Ed.), *Oxford handbook of Latin American social movements* (pp. 143–161). Oxford University Press.

Anguelovski, I., & Alier, J. M. (2014). The 'environmentalism of the poor' revisited: Territory and place in disconnected glocal struggles. *Ecological Economics, 102*, 167–176. https://doi.org/10.1016/j.ecolecon.2014.04.005

Barros, B., & Wilk, R. (2021). The outsized carbon footprints of the super-rich. *Sustainability: Science, Practice and Policy, 17*(1), 316–322. https://doi.org/10.1080/15487733.2021.1949847

Briscoe, M. D., Givens, J. E., & Alder, M. (2021). Intersectional indicators: A race and sex-specific analysis of the carbon intensity of well-being in the United States, 1998–2009. *Social Indicators Research, 155*, 97–116. https://doi.org/10.1007/s11205-021-02613-x

Brulle, R. J. (2021). Networks of opposition: A structural analysis of US climate change countermovement coalitions 1989–2015. *Sociological Inquiry, 91*(3), 603–624. https://doi.org/10.1111/soin.12333

Bullard, R. D. (1983). Solid waste sites and the black Houston community. *Sociological Inquiry, 53*(2–3), 273–288. https://doi.org/10.1111/j.1475-682X.1983.tb00037.x

Bullard, R. D. (1990). *Dumping in dixie: Race, class, and environmental quality*. Westview Press.

Bunker, S. G. (1984). Modes of extraction, unequal exchange, and the progressive underdevelopment of an extreme periphery: The Brazilian Amazon, 1600–1980. *American Journal of Sociology, 89*(5), 1017–1064. https://doi.org/10.1086/227983

Caniglia, B. S., Jorgenson, A., Malin, S. A., Peek, L., & Pellow, D. N. (2021). Introduction: A twenty-first century public environmental sociology. In B. S. Caniglia, A. Jorgenson, S. A. Malin, L. Peek, D. N. Pellow, & X. Huang (Eds.), *Handbook of environmental sociology* (pp. 1–11). Springer Nature.

Carolan, M. S. (2004). Ecological modernization theory: What about consumption? *Society and Natural Resources, 17*(3), 247–260. https://doi.org/10.1080/0894120490270294

Ciplet, D. (2022). Transition coalitions: Toward a theory of transformative just transitions. *Environmental Sociology, 8*(3), 315–330. https://doi.org/10.1080/23251042.2022.2031512

Ciplet, D., & Harrison, J. L. (2020). Transition tensions: Mapping conflicts in movements for a just and sustainable transition. *Environmental Politics, 29*(3), 435–456. https://doi.org/10.1080/09644016.2019.1595883

Clark, B., & Foster, J. B. (2009). Ecological imperialism and the global metabolic rift: Unequal exchange and the guano/nitrates trade. *International Journal of Comparative Sociology, 50*(3–4), 311–334. https://doi.org/10.1177/0020715209105144

Dietz, T., & Jorgenson, A. (2013). *Structural human ecology: New essays in risk, energy, and sustainability*. Washington State University Press.

Dietz, T., & Jorgenson, A. K. (2014). Towards a new view of sustainable development: Human well-being and environmental stress. *Environmental Research Letters, 9*(3), 031001. https://doi.org/10.1088/1748-9326/9/3/031001

Dietz, T., Shwom, R. L., & Whitley, C. T. (2020). Climate change and society. *Annual Review of Sociology, 46*(1), 135–158. https://doi.org/10.1146/annurev-soc-121919-054614

Dietz, T., & York, R. (2021). Structural human ecology. In B. S. Caniglia, A. Jorgenson, S. A. Malin, L. Peek, D. N. Pellow, & X. Huang (Eds.), *Handbook of environmental sociology* (pp. 439–456). Springer Nature.

Downey, L. (2015). *Inequality, democracy, and the environment*. NYU Press.

EJAtlas. Accessed April 8, 2024. Available at: https://ejatlas.org/

Ergas, C., McKinney, L., & Bell, S. E. (2021). Intersectionality and the environment. In B. S. Caniglia, A. Jorgenson, S. A. Malin, L. Peek, D. N. Pellow, & X. Huang (Eds.), *Handbook of environmental sociology* (pp. 15–34). Springer Nature.

Falzon, D., & Sen, R. (2024). A call for a sociology of adaptation. *Sociological Forum, 39*(2), 135–148. https://doi.org/10.1111/socf.12998

Falzon, D., Roberts, J. T., & Brulle, R. J. (2021). Sociology and climate change: A review and research agenda. In Caniglia, B. S., Jorgenson, A., Malin, S. A., Peek, L., Pellow, D. N., & Huang, X. (Eds.), *Handbook of environmental sociology* (pp. 189–217). Springer Nature.

Farrell, J. (2016). Corporate funding and ideological polarization about climate change. *Proceedings of the National Academy of Sciences, 113*(1), 92–97. https://doi.org/10.1073/pnas.1509433112

Fisher, D. R. (2024). *Saving ourselves: From climate shocks to climate action.* Columbia University Press.

Fisher, D. R., & Freudenburg, W. R. (2001). Ecological modernization and its critics: Assessing the past and looking toward the future. *Society & Natural Resources, 14*(8), 701–709. https://doi.org/10.1080/08941920119315

Fitzgerald, J. B. (2022). Working time, inequality and carbon emissions in the United States: A multi-dividend approach to climate change mitigation. *Energy Research & Social Science, 84*, Article 102385. https://doi.org/10.1016/j.erss.2021.102385

Foster, J. B. (1999). Marx's theory of metabolic rift: Classical foundations for environmental sociology. *American Journal of Sociology, 105*(2), 366–405. https://doi.org/10.1086/210315

Foster, J. B., Clark, B., & York, R. (2010). Capitalism and the curse of energy efficiency. *Monthly Review, 62*(6), 1–12. https://monthlyreview.org/articles/capitalism-and-the-curse-of-energy-efficiency/

Foster, J. B., Clark, B., & York, R. (2011). *The ecological rift: Capitalism's war on the earth*. NYU Press.

Givens, J. E. (2015). Urbanization, slums, and the carbon intensity of well-being: Implications for sustainable development. *Human Ecology Review, 22*(1), 107–128. https://www.jstor.org/stable/24875151

Givens, J. E. (2017). World society, world polity, and the carbon intensity of well-being, 1990–2011. *Sociology of Development, 3*(4), 403–435. https://doi.org/10.1525/sod.2017.3.4.403

Givens, J. E. (2018). Ecologically unequal exchange and the carbon intensity of well-being, 1990–2011. *Environmental Sociology, 4*(3), 311–324. https://doi.org/10.1080/23251042.2018.1436878

Givens, J. E., & Huang, X. (2021). Ecologically unequal exchange and environmental load displacement: Global perspectives on structural inequalities and the environment. In B. S. Caniglia, A. Jorgenson, S. A. Malin, L. Peek, D. N. Pellow, & X. Huang (Eds.), *Handbook of environmental sociology* (pp. 53–70). Springer Nature.

Givens, J. E., Huang, X., & Jorgenson, A. K. (2019). Ecologically unequal exchange: A theory of global environmental injustice. *Sociology Compass, 13*(5), Article e12693. https://doi.org/10.1111/soc4.12693

Givens, J. E., Kelly, O. M., & Jorgenson, A. K. (2023). Inequality, emissions, and human well-being. In M. A. Long, M. J. Lynch, & P. B. Stretesky (Eds.), *Handbook on inequality and the environment* (pp. 308–324). Edward Elgar Publishing.

Gould, K. A., Pellow, D. N., & Schnaiberg, A. (2004). Interrogating the treadmill of production: Everything you wanted to know about the treadmill but were afraid to ask. *Organization & Environment, 17*(3), 296–316. https://doi.org/10.1177/1086026604268747

Grant, D., Jorgenson, A., & Longhofer, W. (2020). *Super polluters: Tackling the world's largest sites of climate-disrupting emissions*. Columbia University Press.

Greiner, P. T., & McGee, J. A. (2020). The asymmetry of economic growth and the carbon intensity of well-being. *Environmental Sociology, 6*(1), 95–106. https://doi.org/10.1080/23251042.2019.1675567

Guenther, K. M. (2024). An invitation to bring animals into feminist and queer sociology. *Sociology Compass, 18*(4), Article e13198. https://doi.org/10.1111/soc4.13198

Harrison, J. L. (2019). *From the inside out: The fight for environmental justice within government agencies.* MIT Press.

Hooks, G., Lengefeld, M., & Smith, C. L. (2021). Recasting the treadmills of production and destruction: New theoretical directions. *Sociology of Development, 7*(1), 52–76. https://doi.org/10.1525/sod.2021.7.1.52

Hooks, G., & Smith, C. L. (2004). The treadmill of destruction: National sacrifice areas and Native Americans. *American Sociological Review, 69*(4), 558–575. https://doi.org/10.1177/000312240406900405

Hornborg, A. (1998). Towards an ecological theory of unequal exchange: Articulating world system theory and ecological economics. *Ecological Economics, 25*(1), 127–136. https://doi.org/10.1016/S0921-8009(97)00100-6

Hornborg, A. (2009). Zero-sum world: Challenges in conceptualizing environmental load displacement and ecologically unequal exchange in the world-system. *International Journal of Comparative Sociology, 50*(3–4), 237–262. https://doi.org/10.1177/0020715209105141

Huang, X. (2024a). Harness the co-benefit and avoid the trade-off: The complex relationship between income inequality and carbon dioxide emissions. *Social Forces, 103*(2), 403–428. https://doi.org/10.1093/sf/soae074

Huang, X. (2024b). Not all emissions are created equal: Multidimensionality in nations' greenhouse gas emissions and the affluence/emissions nexus. *Socius, 10*, 23780231241238960. https://doi.org/10.1177/23780231241238962

Hubacek, K., Baiocchi, G., Feng, K., Muñoz Castillo, R., Sun, L., & Xue, J. (2017). Global carbon inequality. *Energy, Ecology and Environment, 2*(6), 361–369. https://doi.org/10.1007/s40974-017-0072-9

Jorgenson, A. K. (2006). Unequal ecological exchange and environmental degradation: A theoretical proposition and cross-national study of deforestation, 1990–2000. *Rural Sociology, 71*(4), 685–712. https://doi.org/10.1526/003601106781262016

Jorgenson, A. K. (2009, March). The sociology of unequal exchange in ecological context: A panel study of lower-income countries, 1975–2000. *Sociological Forum, 24*(1), 22–46. https://doi.org/10.1111/j.1573-7861.2008.01085.x

Jorgenson, A. K. (2015). Inequality and the carbon intensity of human well-being. *Journal of Environmental Studies and Sciences, 5*, 277–282. https://doi.org/10.1007/s13412-015-0234-z

Jorgenson, A. K. (2016a). Environment, development, and ecologically unequal exchange. *Sustainability, 8*(3), 227. https://doi.org/10.3390/su8030227

Jorgenson, A. K. (2016b). The sociology of ecologically unequal exchange, foreign investment dependence and environmental load displacement: Summary of the literature and implications for sustainability. *Journal of Political Ecology, 23*(1), 333–349. https://doi.org/10.2458/v23i1.20221

Jorgenson, A. K., & Clark, B. (2012). Are the economy and the environment decoupling? A comparative international study, 1960–2005. *American Journal of Sociology, 118*(1), 1–44. https://doi.org/10.1086/665990

Jorgenson, A. K., Clark, B., Thombs, R. P., Kentor, J., Givens, J. E., Huang, X., El Tinay, H., Auerbach, D., & Mahutga, M. C. (2023). Guns versus climate: How militarization amplifies the effect of economic growth on carbon emissions. *American Sociological Review, 88*(3), 418–453. https://doi.org/10.1177/00031224231169790

Jorgenson, A. K., & Dietz, T. (2015). Economic growth does not reduce the ecological intensity of human well-being. *Sustain Science, 10*, 149–156. https://doi.org/10.1007/s11625-014-0264-6

Jorgenson, A. K., Dietz, T., & Kelly, O. (2018). Inequality, poverty, and the carbon intensity of human well-being in the United States: A sex-specific analysis. *Sustainability Science, 13*, 1167–1174. https://doi.org/10.1007/s11625-017-0517-2

Jorgenson, A., El Tinay, H., Fitzgerald, J., Givens, J., Goh, T., Huang, X., Kelly, O., Rieger, A., & Thombs, R. P. (2024). Advances in research on anthropogenic drivers of climate change. In S. R. Brechin & S. Lee (Eds.), *Routledge handbook of climate change and society* (pp. 60–76). Routledge.

Jorgenson, A. K., Fiske, S., Hubacek, K., Li, J., McGovern, T., Rick, T., Schor, J. B., Solecki, W., York, R., & Zycherman, A. (2019). Social science perspectives on drivers of and responses to global climate change. *Wiley Interdisciplinary Reviews: Climate Change, 10*(1), Article e554. https://doi.org/10.1002/wcc.554

Jorgenson, A. K., & Givens, J. E. (2014). The emergence of new world-systems perspectives on global environmental change. In S. Lockie, D. A. Sonnenfeld, & D. R. Fisher (Eds.), *Routledge international handbook of social and environmental change* (pp. 31–44). Routledge.

Jorgenson, A. K., & Givens, J. (2015). The changing effect of economic development on the consumption-based carbon intensity of well-being, 1990–2008. *PLOS One, 10*(5), Article e0123920. https://doi.org/10.1371/journal.pone.0123920

Jorgenson, A., Schor, J., & Huang, X. (2017). Income inequality and carbon emissions in the United States: A state-level analysis, 1997–2012. *Ecological Economics, 134*, 40–48. https://doi.org/10.1016/j.ecolecon.2016.12.016

Kallis, G. (2011). In defence of degrowth. *Ecological Economics, 70*(5), 873–880. https://doi.org/10.1016/j.ecolecon.2010.12.007

Kallis, G., Kostakis, V., Lange, S., Muraca, B., Paulson, S., & Schmelzer, M. (2018). Research on degrowth. *Annual Review of Environment and Resources, 43*(1), 291–316. https://doi.org/10.1146/annurev-environ-102017-025941

Klinenberg, E., Araos, M., & Koslov, L. (2020). Sociology and the climate crisis. *Annual Review of Sociology, 46*(1), 649–669. https://doi.org/10.1146/annurev-soc-121919-054750

Klinsky, S., Roberts, T., Huq, S., Okereke, C., Newell, P., Dauvergne, P., O'Brien, K., Schroeder, H., Tschakert, P., Clapp, J., & Bauer, S. (2017). Why equity is fundamental in climate change policy research. *Global Environmental Change, 44*, 170–173. https://doi.org/10.1016/j.gloenvcha.2016.08.002

Knight, K. W., Schor, J. B., & Jorgenson, A. K. (2017). Wealth inequality and carbon emissions in high-income countries. *Social Currents, 4*(5), 403–412. https://doi.org/10.1177/2329496517704872

Kojola, E., & Agyeman, J. (2021). Just transitions and labor. In B. S. Caniglia, A. Jorgenson, S. A. Malin, L. Peek, D. N. Pellow, & X. Huang (Eds.), *Handbook of environmental sociology* (pp. 115–138). Springer Nature.

Ladegaard, I., & Rieger, A. (2024). How "ceremonial openness" Prevents organizational change: An analysis of corporate earnings calls in the oil and gas industry, 2007–2020. *Social Problems*, spae045. https://doi.org/10.1093/socpro/spae045

Liévanos, R. S., Wilder, E., Richter, L., Carrera, J., & Mascarenhas, M. (2021). Challenging the white spaces of environmental sociology. *Environmental Sociology, 7*(2), 103–109. https://doi.org/10.1080/23251042.2021.1902665

Malin, S. A., & Ryder, S. S. (2018). Developing deeply intersectional environmental justice scholarship. *Environmental Sociology, 4*(1), 1–7. https://doi.org/10.1080/23251042.2018.1446711

Markkanen, S., & Anger-Kraavi, A. (2019). Social impacts of climate change mitigation policies and their implications for inequality. *Climate Policy, 19*(7), 827–844. https://doi.org/10.1080/14693062.2019.1596873

Martinez-Alier, J., Temper, L., Del Bene, D., & Scheidel, A. (2016). Is there a global environmental justice movement? *The Journal of Peasant Studies, 43*(3), 731–755. https://doi.org/10.1080/03066150.2016.1141198

Maung, R., & Pellow, D. N. (2021). Environmental justice. In B. S. Caniglia, A. Jorgenson, S. A. Malin, L. Peek, D. N. Pellow, & X. Huang (Eds.), *Handbook of environmental sociology* (pp. 35–52). Springer Nature.

McClintock, N. (2010). Why farm the city? Theorizing urban agriculture through a lens of metabolic rift. *Cambridge Journal of Regions, Economy and Society, 3*(2), 191–207. https://doi.org/10.1093/cjres/rsq005

McGee, J. A., Ergas, C., Greiner, P. T., & Clement, M. T. (2017). How do slums change the relationship between urbanization and the carbon intensity of well-being? *PLOS One, 12*(12), Article e0189024. https://doi.org/10.1371/journal.pone.0189024

McGee, J. A., Greiner, P. T., Christensen, M., Ergas, C., & Clement, M. T. (2020). Gender inequality, reproductive justice, and decoupling economic growth and emissions: A panel analysis of the moderating association of gender equality on the relationship between economic growth and CO2 emissions. *Environmental Sociology, 6*(3), 254–267. https://doi.org/10.1080/23251042.2020.1736364

Mohai, P., Pellow, D., & Roberts, J. T. (2009). Environmental justice. *Annual Review of Environment and Resources, 34*(1), 405–430. https://doi.org/10.1146/annurev-environ-082508-094348

Muradian, R., O'Connor, M., & Martinez-Alier, J. (2002). Embodied pollution in trade: Estimating the 'environmental load displacement' of industrialised countries. *Ecological Economics, 41*(1), 51–67. https://doi.org/10.1016/S0921-8009(01)00281-6

Pellow, D. N. (2017). *What is critical environmental justice?* Wiley.

Piketty, T. (2014a). *Capital in the twenty-first century*. Harvard University Press.

Piketty, T. (2014b). Capital in the twenty-first century: A multidimensional approach to the history of capital and social classes. *British Journal of Sociology, 65*(4), 736–747. https://doi.org/10.1111/1468-4446.12115

Purucker, D. (2021). Critical environmental justice and the state: A critique of pellow. *Environmental Sociology, 7*(3), 176–186. https://doi.org/10.1080/23251042.2021.1878575

Rieger, A. (2024). Are corporations responding to civil society pressure? A multi-level analysis of corporate emissions. *Sociology of Development, 10*(3), 310–334. https://doi.org/10.1525/sod.2023.0047

Rieger, A., & Schor, J. B. (2021). Consumption. In B. S. Caniglia, A. Jorgenson, S. A. Malin, L. Peek, D. N. Pellow, & X. Huang (Eds.), *Handbook of environmental sociology* (pp. 71–87). Springer Nature.

Roberts, J. T. (2001). Global inequality and climate change. *Society & Natural Resources, 14*(6), 501–509. https://doi.org/10.1080/08941920118490

Roberts, J. T., & Parks, B. (2006). *A climate of injustice: Global inequality, north–south politics, and climate policy*. MIT Press.

Roberts, J. T., & Parks, B. C. (2009). Ecologically unequal exchange, ecological debt, and climate justice: The history and implications of three related ideas for a new social movement. *International Journal of Comparative Sociology, 50*(3–4), 385–409. https://doi.org/10.1177/0020715209105147

Rosa, E., & Dietz, T. (2012). Human drivers of national greenhouse-gas emissions. *Nature Climate Change, 2*, 581–586. https://doi.org/10.1038/nclimate1506

Ryder, S. S. (2018). Developing an intersectionally-informed, multi-sited, critical policy ethnography to examine power and procedural justice in multiscalar energy and climate change decisionmaking processes. *Energy Research & Social Science, 45*, 266–275. https://doi.org/10.1016/j.erss.2018.08.005

Schlosberg, D. (2013). Theorising environmental justice: The expanding sphere of a discourse. *Environmental Politics, 22*(1), 37–55. https://doi.org/10.1080/09644016.2013.755387

Schnaiberg, A. (1980). *The environment: From surplus to scarcity*. Oxford University Press.

Schor, J. B., & Jorgenson, A. K. (2019). Is it too late for growth? *Review of Radical Political Economics, 51*(2), 320–329. https://doi.org/10.1177/0486613419831109

Smith, C. L., Hooks, G., & Lengefeld, M. (2020). Treadmills of production and destruction in the anthropocene: Coca production and gold mining in Colombia and Peru. *Journal of World-Systems Research, 26*(2), 231–262. https://doi.org/10.5195/jwsr.2020.981

Spaargaren, G., & Mol, A. P. (1992). Sociology, environment, and modernity: Ecological modernization as a theory of social change. *Society & Natural Resources, 5*(4), 323–344. https://doi.org/10.1080/08941929209380797

Stoddard, I., Anderson, K., Capstick, S., Carton, W., Depledge, J., Facer, K., Gough, C., Hache, F., Hoolohan, C., Hultman, M., Hällström, N., Kartha, S., Klinsky, S., Kuchler, M., Lövbrand, E., Nasiritousi, N., Newell, P., Peters, G. P., Sokona, Y., … Williams, M. (2021). Three decades of climate mitigation: why haven't we bent the global emissions curve? *Annual Review of Environment and Resources*, 46(1), 653–689. https://doi.org/10.1146/annurev-environ-012220-011104

Taylor, N., Sutton, Z., & Wilkie, R. (2018). A sociology of multi-species relations. *Journal of Sociology, 54*(4), 463–466. https://doi.org/10.1177/1440783318816214

Temper, L., Demaria, F., Scheidel, A., Del Bene, D., & Martinez-Alier, J. (2018). The Global Environmental Justice Atlas (EJAtlas): Ecological distribution conflicts as forces for sustainability. *Sustainability Science, 13*(3), 573–584. https://doi.org/10.1007/s11625-018-0563-4

Thombs, R. P., Huang, X., & Fitzgerald, J. B. (2022). What goes up might not come down: Modeling directional asymmetry with large-N, large-T data. *Sociological Methodology, 52*(1), 1–29. https://doi.org/10.1177/00811750211046307

Wang, S., Xie, Z., Wu, R., & Feng, K. (2022). How does urbanization affect the carbon intensity of human well-being? A global assessment. *Applied Energy, 312*, 118798. https://doi.org/10.1016/j.apenergy.2022.118798

Warlenius, R. (2018). Decolonizing the atmosphere: The climate justice movement on climate debt. *The Journal of Environment & Development, 27*(2), 131–155. https://doi.org/10.1177/1070496517744593

Wittman, H. (2009). Reworking the metabolic rift: La Vía Campesina, agrarian citizenship, and food sovereignty. *The Journal of Peasant Studies, 36*(4), 805–826. https://doi.org/10.1080/03066150903353991

Whyte, K. (2020). Indigenous environmental justice: Anti-colonial action through kinship. In B. Brendan Coolsaet (Ed.), *Environmental justice* (pp. 266–278). Routledge.

York, R., & Rosa, E. A. (2003). Key challenges to ecological modernization theory: Institutional efficacy, case study evidence, units of analysis, and the pace of eco-efficiency. *Organization & Environment, 16*(3), 273–288. https://doi.org/10.1177/1086026603256299

York, R., Rosa, E. A., & Dietz, T. (2003). STIRPAT, IPAT and ImPACT: Analytic tools for unpacking the driving forces of environmental impacts. *Ecological Economics, 46*(3), 351–365. https://doi.org/10.1016/S0921-8009(03)00188-5

York, R., Rosa, E. A., & Dietz, T. (2010). Ecological modernization theory: Theoretical and empirical challenges. In M. R. Redclift (Ed.), *The international handbook of environmental sociology* (pp. 77–90). Edward Elgar.

4

Intersectionality

Ana Terra Amorim-Maia and Isabelle Anguelovski

Introduction

Across the globe, societies pursuing just transformations find themselves grappling with compounding crises and intersecting inequalities. These include escalating gender-based and racially motivated violence, deepening socio-economic inequality, strained care systems, and the widespread impacts of environmental degradation and climate change. Addressing these challenges individually risks privileging certain groups at the expense of others, thereby intensifying existing social, economic, and political divides and diverting attention from the broader need to pursue more cohesive and transformative strategies. In this context, it is increasingly imperative for policymakers and planners across scales—from local to global—to adopt approaches that

A. T. Amorim-Maia (✉)
BC3 Basque Centre for Climate Change, Bilbao, Spain
e-mail: ana.amorim@bc3research.org

I. Anguelovski
Institute of Environmental Science and Technology (ICTA-UAB), Universitat Autonoma de Barcelona, Bellaterra, Spain

Catalan Institution for Research and Advanced Studies, ICREA, Barcelona, Spain

I. Anguelovski
e-mail: isabelle.anguelovski@uab.cat

D. J. Davidson et al. (eds.), *Environment and Society*,
https://doi.org/10.1007/978-3-032-27059-7_4

not only help illuminate intersecting inequalities, but also enable more integrated solutions to address them, fostering pathways toward transformative futures.

This chapter introduces the concept of *intersectionality* as a guiding framework for just transformations. Rooted in Critical Race Theory and Black Feminist scholarship, intersectionality examines how interconnected social identities—such as gender, race[4], class, age, disability, and sexual orientation—shape experiences of privilege, discrimination, and oppression (Cho et al., 2013; Crenshaw, 1989). It poses that approaches that focus on a single axis of inequality (e.g., gender *or* race) often fall short of addressing the complex, overlapping challenges that arise at their intersection, including marginalisation, climate vulnerability, and environmental injustice (Anguelovski et al., 2020; Timmons Roberts et al., 2018). For instance, while certain policies address gender equity—such as initiatives to advance women's rights in labour markets or health systems—or tackle racial injustice—such as anti-discrimination policies in education or housing—these efforts are frequently siloed. In some cases, single-axes policies may even exacerbate inequalities at the intersections that they overlook—for example, racial equity initiatives that primarily benefit men of colour can reinforce gender disparities among women of colours, or gender-focused programs that centre white women's experiences may marginalise racialised or migrant women. Consequently, these policies provide fragmented and incomplete responses to systemic challenges, where individuals experience compounded harms precisely because of how these systems of inequality interact.

We frame intersectionality as an organising principle for just transformations in the context of the accelerating environmental and climate crises. This approach enables socially transformative frameworks that address the root causes of interconnected inequalities, particularly those related to environmental and climate justice. An intersectional approach helps decision-makers critically examine how power and privilege operate across society, enabling them to more effectively recognise and address cumulative and overlapping inequalities.. Embedding intersectionality into planning and policymaking can thus support efforts to challenge the entrenched power dynamics that perpetuate environmental and social vulnerabilities (Amorim-Maia et al., 2022; Phoenix & Pattynama, 2006).

Although the academic community increasingly acknowledges the relevance of intersectional approaches to environmental justice and climate action (Djoudi et al., 2016; Kaijser & Kronsell, 2014), greater attention is still needed on how to apply these approaches in practice. In response, this chapter offers a conceptual framework for intersectional justice in the context of socioenvironmental transformations, organised around five core principles, first developed by Amorim-Maia et al. (2022). These are illustrated through examples from Barcelona, a city recognised for its innovative efforts to integrate intersectionality into environmental governance and urban planning.

Between 2015 and 2023, Barcelona was led by Mayor Ada Colau, whose administration brought a bold vision of feminist, inclusive, and climate-conscious urban governance. Under her leadership, the city framed itself as a *caring city*, embedding ethics of care, participation, and social justice across policy domains. This vision translated into pioneering initiatives such as the VilaVeina (local care hubs), climate shelters, and neighbourhood-scale "Superblocks" that aimed to reduce car dominance while greening the city. Policies during this period sought to integrate gender, class, and care responsibilities into traditionally siloed areas like mobility, housing, and climate action. For instance, urban mobility plans were reoriented to support everyday care-related journeys, and public housing policies were adapted to prioritise women and single-parent households. These initiatives were supported by institutional mechanisms such as the Department of Gender Mainstreaming and the Barcelona Cuida programme for carers. Although not always framed explicitly as "intersectional", many of these efforts aligned with the principles of intersectional thinking by addressing overlapping forms of inequality and systemic exclusion. While the long-term institutionalisation of these policies remains uncertain following political changes in 2023, Barcelona's experience provides valuable insights into potential pathways for operationalising intersectional approaches to just transitions. We conclude by reflecting on the broader implications of intersectionality for advancing just transformations and proposing avenues for future research.

Intersectionality and its Application in Environmental and Climate Studies

Intersectionality provides a critical lens for understanding how multiple forms of social inequality and vulnerability intersect and reinforce one another. While originally developed within Black Feminist Thought, it has since been applied across diverse fields—from gender studies to psychology and organisational studies—and is increasingly gaining attention in environmental studies—particularly environmental justice (Ducre, 2018; Malin & Ryder, 2018; Malin et al., 2019). Intersectionality has also been applied to climate mitigation efforts, including research on energy transitions (Cannon & Chu,

2021; Johnson et al., 2020) and land restoration (Thompson-Hall, 2016). Moreover, it is increasingly applied to climate adaptation research, helping reveal the interconnected systems of oppression that restrict adaptive capacity and exacerbate vulnerabilities (Djoudi et al., 2016; Kaijser & Kronsell, 2014). While much of the research focuses on rural settings, such as women's land access in South Asia or Ghana (Ravera et al., 2016; Sultana, 2014), attention to urban applications is growing (e.g., Sultana, 2020; Wilson & Chu, 2019). For example, researchers on urban greening have been calling for intersectional approaches to uncover the multiple forms of inequalities that can result from green infrastructure projects, especially as those are often embedded in green growth, urban green grabbing, and racially exclusionary logics (Anguelovski et al., 2020). However, the application of intersectionality in both climate and environmental governance remains fragmented, and a more systematic integration of intersectional thinking is necessary, as traditional environmental governance and climate action have often been criticised for perpetuating structural vulnerabilities through technocratic and exclusionary approaches that overlook historical patterns of inequality and deepen social and environmental disparities (Chu et al., 2017; Robin & Castán Broto, 2020). In response, an intersectional approach seeks to address interconnected socioenvironmental injustices, challenging entrenched power dynamics and fostering more inclusive transformations.

To address this gap, Amorim-Maia et al. (2022) argued for a intersectional framework structured around five key principles for advancing justice-oriented transformations in environment-society relations. In this framework, intersectionality moves beyond individual identity categories or drivers of inequality (e.g., racism *or* sexism) to address the intersecting, structural systems of power that create and perpetuate discrimination and oppression (Cho et al., 2013; McCall, 2005). By examining how intersecting identities shape experiences of privilege and disadvantage, intersectionality enables scholars and practitioners to situate these dynamics within their historical and spatial contexts (Anguelovski et al., 2020). Furthermore, by focusing on the distribution of care needs and responsibilities, intersectionality critiques historical—mostly paternalistic, misogynist, and white—hierarchies that dictate who provides and receives care (Hankivsky, 2014; Raghuram, 2016), exposing gendered labour inequalities and acknowledging reproductive work as essential to societal well-being. Intersectionality also highlights the role of place and space in shaping experiences of privilege and oppression, underscoring the importance of context in understanding systemic inequality (Davis, 2008). Rather than a universal framework, intersectionality is a situated concept that highlights how inequality and privilege are experienced differently depending on geographic, historical, and social context. Thus, it draws attention both to the daily, embodied experiences of exclusion faced by marginalised groups and to the specific conditions in which these experiences unfold. At its core, intersectionality seeks not only to analyse but also to challenge dynamics of power and privilege, fostering cross-identity forms of activism and coalitions that can drive justice-oriented transformations (Collins & Bilge, 2020).

Intersectionality in the Context of Socioenvironmental Transformations: A Conceptual Framework

This section introduces a conceptual framework structured around five core principles for integrating intersectionality into environmental and climate governance to advance socioenvironmental transformations. These components highlight the need to:

1. Address the structures and systems that perpetuate racial and gender inequalities;
2. Redress differential vulnerabilities across social groups and contexts;
3. Centre ethics and politics of care in governance and decision-making processes;
4. Embrace place-based approaches, including those that acknowledge marginalised forms of place-making; and
5. Foster cross-identity activism and community resilience.

Each subcomponent is explored through a theoretical lens and critiqued, with examples of transformative policies and initiatives from Barcelona. These examples illustrate how intersectionality can be operationalised to address systemic inequalities in environmental, climate, and urban planning, guiding efforts toward more inclusive and just transformations. A visual representation of the framework is provided in Fig. 4.1.

Fig. 4.1 Intersectional justice in the context of socioenvironmental transformations: A Conceptual Framework. Reproduced from: Amorim-Maia, A. T., Anguelovski, I., Chu, E., & Connolly, J. (2022)

Addressing Systemic Drivers of Racial and Gender Inequalities

At its core, intersectionality emerged to address how the struggles of women of colour often fall between the gaps of feminist and anti-racist discourse (Crenshaw, 1989). In her seminal work, legal scholar and civil rights advocate Kimberlé Crenshaw introduced the concept to illustrate how U.S. legal and institutional frameworks failed to recognise discrimination occurring simultaneously on the basis of race and gender. While intersectionality is not limited to these two axes and context may centre other dimensions, its origins lie in highlighting the specific, compounded marginalisation experience of women of colour. The framework proposed here acknowledges that history, not to prioritise certain axes of oppression or regional epistemologies, but to acknowledge how capitalism, colonialism, and patriarchy have historically and persistently excluded racialised and gendered bodies. In this light, the first principle of the intersectional framework highlights the need to recognise and challenge the systemic structures that reinforce racial and gender inequalities, with attention to how historic patterns of exclusion continue to shape socioenvironmental systems (Pulido, 2017; Robinson, 2000).

Advancing intersectionality in environmental and climate governance requires moving beyond traditional Marxist critiques of capitalism, which often focus solely on class-based exploitation, and instead acknowledging how capitalism has historically relied on the oppression of racial minorities and the undervaluation of women's labour. Intersectionality helps illuminate how race- and gender-based exclusionary practices drive systemic inequalities, and how these patterns of devaluation and exclusion create compounded vulnerabilities that disproportionately expose women, racial minorities, and other marginalised groups to climate and environmental risks. For instance, it is well documented that racialised women frequently encounter systemic barriers to accessing land, resources, education, institutional support, and decision-making processes, further exacerbating their precarity and vulnerability (Federici, 2004; Virdee, 2019).

However, it is also critical to avoid reductive narratives that portray marginalised groups, especially women and Indigenous communities, as either passive victims or inherent leaders in resilience and environmental stewardship. Such narratives risk reinforcing stereotypes and overburdening these groups with additional responsibilities. Instead, an intersectional approach critiques the power dynamics and structural inequities that perpetuate social and environmental injustices, ensuring that policies prioritise the voices, perspectives, and needs of historically marginalised communities (Enarson, 2013; Resurrección, 2017).

Environmental politics and governance remain some of the core domains of social and environmental injustices. Despite being most affected by climate and environmental impacts, women and racial minorities remain underrepresented in decision-making processes related to climate and environmental governance, which are often dominated by technocratic, expert-driven approaches. This lack of representation frequently results in policies that are blind to race and gender, leaving systemic inequalities unchallenged and creating new tensions over time. To counter this, intersectional justice calls for explicitly feminist and anti-racist approaches to policymaking that actively critique and dismantle the oppressive systems that sustain inequalities (Anguelovski et al., 2019).

Examples of incorporating feminist or anti-racist perspectives into governance are numerous, though integrating both principles simultaneously remains relatively uncommon. Barcelona embedded a feminist perspective across municipal governance to tackle entrenched gender inequalities, ensuring gender mainstreaming in all sectors and policies. This included environmental and climate initiatives like the Superblocks, climate shelters, and energy poverty reduction programs, all of which had a gender perspective guided by the Barcelona II Plan for Gender Justice. Other foundational feminist policymaking frameworks also guided several of the city's governance and planning efforts, emphasising proximity, access, care, and inclusion in the deployment of different urban projects. A notable anti-racist initiative is Barcelona's Office for Non-Discrimination, which works to combat racial and ethnic discrimination through legal support, mediation, and educational campaigns to promote social equality, awareness, and cohesion in the city. Although commendable on their own, efforts to address intersecting inequalities—such as those related to gender, race, ethnicity, and social class simultaneously—remain in their early stages (Amorim-Maia et al., 2024).

Redressing Differential Vulnerabilities

Beyond gender and race, intersectionality contends with broader systems of oppression that shape how vulnerability is produced and experienced—in this case, socioenvironmental and climate vulnerability, emphasising the interplay between structural and political inequalities as fundamental drivers of vulnerability. Thus, an intersectional approach helps recognise and redress differential vulnerabilities and their drivers — the ways intersecting social inequalities and the root processes behind them exacerbate exposure to climate and environmental risks. Differential vulnerability, first developed in feminist scholarship, highlights how factors like gender, race, ethnicity, age, and religion interact with cultural and social norms to produce diverse experiences of vulnerability (Blaikie et al., 1994; Cho et al., 2013).

Rather than viewing vulnerability as static or population-specific, intersectionality frames it as a dynamic function of compounding risks, such as natural disasters, extreme climate events, and land conflicts, and structural drivers of inequality. An intersectional perspective deepens understanding of how these risks overlap with systemic inequalities to shape unique vulnerabilities and responses in various contexts. While research on strategies to address differential vulnerabilities is growing, limited attention has been given to redressing their root causes, particularly in systemic or multi-scale contexts.

Vulnerabilities are often compounded by pre-existing challenges such as inadequate infrastructure, insecure land tenure, and limited political representation and rights, which can amplify environmental risk for marginalised populations. These vulnerabilities have been further compounded by global crises like the COVID-19 pandemic, which revealed and intensified disparities in exposure and resilience (Berrouet et al., 2017; Cardona et al., 2012; Kuran et al., 2020; Pelling et al., 2021).

The increasing reliance on privatised and market-based climate finance poses significant challenges to addressing differential vulnerabilities equitably. For instance, studies have shown that adaptation funding often prioritises private investments, leading to inequitable outcomes, such as land seizures, displacement, and "resilience gentrification", where climate adaptation projects tend to raise property values and displace low-income residents. Redressing differential vulnerabilities requires countering such trends by prioritising policies that improve access to green spaces while regulating private investment, working toward land ownership and control for non-white groups, and ensuring housing stability and rights for lower-income and -wealth residents (BCNUEJ, 2021; Connolly, 2018).

Barcelona took steps to reduce differential vulnerabilities, integrating equitable greening and housing rights. The city implemented regulations on short-term rentals, mandated a 30% social housing quota in new developments (with at least 20% allocated to women and single-parent families), retrofitted vulnerable buildings, and expanded green infrastructure in underserved neighbourhoods. These efforts were part of the Neighbourhood Plan (Pla de de Barris) and the "Urban Regeneration Program", which aimed to reduce vulnerability by enhancing housing, green spaces, and public areas for leisure and recreation. Despite these efforts, concerns over rising housing prices and green gentrification persisted. To counteract speculation, the city linked the creation of new green spaces to a Land Use Plan regulating the housing market and commercial activity, and adapted the Superblocks model from a 9 × 9 block layout to a Green Axes approach to distribute the benefits of greening more evenly. Another key initiative was the creation of a "network of climate shelters", which connects public facilities and green infrastructure to provide shaded, safe spaces for shelter during extreme heat. The initiative started at the city-level and has now expanded to 35 municipalities in the metropolitan area, offering over 500 spaces of heat and cold refuge. While not a panacea for extreme temperatures, these shelters offer essential indoor and outdoor public spaces that also help combat social isolation and support health and well-being (Amorim-Maia et al., 2023, 2024).

Prioritising Politics and Ethics of Care

Intersectionality not only exposes entrenched social inequities but also reveals their manifestation in care inequities — disparities in the distribution and recognition of care work, which is fundamental to human survival. In many capitalist societies, care has been feminised and privatised, with reproductive and unpaid caregiving roles largely relegated to women, particularly women of colour, migrants, and working-class individuals. These roles often go undervalued and are exploited to sustain economic systems while perpetuating gendered and racialised hierarchies (Ehrenreich & Hochschild, 2004; Glenn, 2012; Held, 2006; Raghuram, 2016).

In environmental governance and climate action, care ethics offer a feminist perspective that challenges traditional technocratic approaches, which often prioritise individual productivity and economic objectives over well-being, trust, and solidarity. Instead, this framework calls for recognising care as a form of essential labour for the protection of socioenvironmental systems and restructuring policies and systems to distribute it more equitably. In cities, this approach involves rethinking planning and governance through a care perspective, creating public policies and services that prioritise the regeneration of physical and emotional well-being, allow for self-care, and create spaces where social equity and participatory governance are transversal objectives of all actions (Adger et al., 2017; Lawson, 2007).

As a self-proclaimed "Caring City", Barcelona has integrated care into urban planning, centring its design and functioning around everyday reproductive needs. The city pioneered the institutionalisation of care ethics in governance through the Government Bill for Democratizing Care Work, which introduced 68 initiatives to promote fairer care distribution, recognise care work as a fundamental right, and address the unequal burdens of caregiving. A key initiative within this framework was the VilaVeïna Project, which created "Care Superblocks" offering resources such as emotional support groups, shared parenting spaces, and legal advice for caregivers. Notably, the city sought to align these Care Superblocks with the broader Superblocks, integrating environmental planning—such as greening and traffic calming—with a feminist care perspective, including expanding walking and cycling networks to prioritise everyday caregiving activities. Beyond infrastructure, Barcelona has implemented policies and programs to support caregivers, recognising their often invisible and undervalued labour. The "Carer Card" (introduced in 2022 through the Barcelona Cuida Program) provided training, legal support, and well-being programs, including assistance for immigrant domestic and care workers to help regularise employment. These initiatives demonstrate how urban planning can be linked to social justice by addressing the feminisation and invisibility of care work and promoting a more democratic and sustainable model of care.

Although these initiatives demonstrate progress, they remain siloed from environmental and climate-related policies and have not always perdured or been fully institutionalised. Integrating environmental considerations and institutional stability into these programs could enhance their impact, emphasising the interconnectedness of care, well-being, and sustainability. In that regard, if green space planners were able to integrate nature's role in emotional and physical health, they could develop green spaces more effectively designed for caregiving, self-care, and well-being. This could include facilities for different vulnerable groups, culturally-inclusive design elements, and tailored social and physical activities based on diverse users' demands and needs. These examples highlight the transformative potential of embedding care ethics into governance to address social inequities and promote resilience (Fisher & Tronto, 1990; Kabisch et al., 2017).

Adopting Place-Based and Place-Making Approaches

Intersectionality emphasises the situatedness of knowledge and the structural role of place and space in shaping experiences of privilege, oppression, and identity. Thus, an intersectional perspective calls for planning and governance approaches centred on local communities, prioritising their relationships with and meanings of place. Place-based approaches shift away from technocratic, expert-driven models and instead foreground the knowledge, needs, and priorities of historically marginalised groups, such as women, racial minorities, and immigrants (Anguelovski et al., 2020; Crenshaw, 1991; Haraway, 1988).

Place-based and place-making approaches challenge dominant planning paradigms often designed by and for privileged social groups. By promoting community ownership of co-designed and co-managed spaces, these approaches foster inclusivity and recognition of diverse histories and everyday lived experiences, including those of extraction, loss, and displacement in view of unequal development. Moreover, they engage with racialised geographies shaped by colonialism, slavery, and systemic racism, recognising subaltern practices of place-making often excluded from formal planning schemes (Escobar, 2001; McKittrick, 2011; Olazabal et al., 2021).

To redress historical and ongoing inequities, this component emphasises actions such as integrating decolonial and inclusive planning approaches, including rights to place and return; confronting the legacy of policed states and disciplined landscapes that constrain marginalised practices and uses of space; and valuing traditional, Indigenous, and local knowledge systems while reflecting on how knowledge is produced and whose voices are amplified. This involves not only recognising Indigenous peoples as knowledge-holders but also embracing Indigenous cosmologies, epistemologies, values, and ways of relating to land, time, and community. These perspectives challenge dominant Western paradigms by offering alternative understandings of care, stewardship, and interdependence. For example, Buen Vivir ("Good Living"), rooted in Andean Indigenous worldviews and feminist and environmental philosophies, offers an ecologically balanced and community-centric vision of development. Its principles have influenced global frameworks, including Barcelona's Global Justice Cooperation Plan (2018–2021), and Ecuador's National Plan for Buen Vivir. Furthermore, in Portland, Oregon, the North/Northeast Preference Policy (colloquially known as the Right to Return policy) is an initiative designed to address the displacement of residents—particularly African American communities—caused by past urban renewal projects in 1960s, with priority given to individuals and families who can demonstrate generational ties to the neighbourhoods affected by urban renewal. These residents can be considered as part of environmental injustice communities, for having often been displaced to areas near industrial zones or highways, exposing them to pollution and other environmental hazards. Such policies address past processes of economic extraction and racism, reparative justice needs, and future community resilience.

In Barcelona, examples of place-making approaches include the City's efforts to expand and diversify participation in environmental politics, acknowledging that simply providing participatory spaces is insufficient for truly representative and meaningful engagement. Thus, the city is actively designing strategies to broaden civic engagement, reaching out to marginalised groups through trusted community figures, providing translation services and accessibility support, adapting meeting locations, and offering childcare and financial remuneration for participation. For instance, the new generation of Superblocks integrated participatory urban innovation, where neighbourhood steering groups collaborated on street-calming and greening projects, allowing residents to co-create and rank proposals before seeking broader community feedback. A further key initiative was the Citizen's Assembly, which engaged 100 residents—selected to represent Barcelona's demographic diversity—in deliberating climate issues. The

Assembly produced 34 proposals to reform energy, mobility, and consumption models, illustrating the transformative potential of intersectional approaches to address socioenvironmental challenges more inclusively. Moreover, place-based approaches have been integrated into the Climate Shelters in Schools initiative, launched in 2019. This project transformed schoolyards vulnerable to heat into "cool islands" through a mix of green (vegetation and shading), blue (water points), and grey (insulation and ventilation) measures (Ruiz-Mallén et al., 2023). Students, educators, and experts co-designed the spaces to address local needs, while incorporating climate education and community participation.

Promoting Cross-Identity and Cross-Vulnerability Climate Action and Community Resilience

Intersectionality inherently involves analysing power relations and structures of subordination, while encouraging collective responses to social injustices and fostering activism. It thus emphasises the need for cross-identity and cross-vulnerability activism, alongside community-driven resilience. This refers to the importance of building alliances across different social movements, identities, and experiences of marginalisation—for instance, recognising how labour precarity intersects with racial and gender discrimination. Intersectionality calls for moving beyond siloed struggles and working in solidarity, thus acknowledging that liberation from one form of oppression requires challenging others. This coalition-based activism can more effectively challenge the drivers of socioenvironmental injustice and promote just and resilient futures within planetary boundaries (Collins & Bilge, 2020; Di Chiro, 2008; Terry, 2009; Thomas, 2022).

By adopting a flexible definition of resilience—communities' ability to adapt, transform, and maintain desired functions in the face of disturbances—this component prioritises mechanisms that empower marginalised populations to address vulnerabilities across social identities. Community-focused experiments, which serve as "purposive interventions" to foster innovation and learning, and governance practices, are one such mechanism. However, these experiments should prioritise community ownership of co-designed and co-managed spaces to ensure equitable and lasting outcomes (Bulkeley et al., 2014; Caprotti & Cowley, 2017).

For highly vulnerable populations, such as the urban poor, displaced persons, or residents of informal settlements, insurgent planning offers a pathway to reshape power dynamics in decision-making processes. This approach supports "citizenship from below", encouraging communities to create their own opportunities and terms of engagement, moving beyond tokenistic invitations to participate. To achieve meaningful community engagement, these efforts must strengthen the representation of minority-focused organisations, improve access to climate and environmental data and the capacity to interpret and apply it effectively, and foster collaborations between local governments, not-for-profit organisations, and cooperative enterprises. Additionally, social media platforms and online hubs serve as valuable tools for advancing intersectional activism. (Chu et al., 2019; Cornwall, 2002).

Barcelona has made strides in promoting citizen-driven environmental projects. One notable initiative was Pla BUITS ("Vacant Lots Plan"), which revitalised disused urban land for temporary community use. During the early 2010s, neighbourhood associations and cooperatives transformed vacant lots into urban gardens, incorporating permaculture, rainwater collection, and environmental education. These spaces not only fostered social cohesion but also provided platforms for community innovation in sustainable practices. Another initiative, the Horts al Terrat green roof project, repurposed unused municipal rooftops into gardens. Led by local non-profits, these spaces serve as therapeutic tools for people with disabilities, the unemployed, and older residents and some have even won international prizes. By improving physical and emotional well-being, fostering inclusion, and offering decision-making autonomy, these projects highlight the intersectional potential of community-driven projects that improve existing land uses while supporting sustainability goals (Triguero-Mas et al., 2020).

Toward a Reflexive and Expansive Use of Intersectionality

While the framework proposed in this chapter offers an applied pathway to embed intersectional thinking in just transformations, it is not without limitations. Intersectionality is not a catch-all solution and, if applied superficially or decontextualised, risks being diluted or reduced to identity-mapping exercises. Crucially, intersectionality is not only about identifying marginalisation but also about interrogating the systems of privilege, power, and normativity that structure those inequalities.

In the context of environmental and climate governance, this means turning the lens toward dominant groups—including white, affluent, cisgender men—and critically examining how their privileges are maintained. It also means interrogating the assumptions underpinning technocratic decision-making, the socio-political profiles of climate denial, and the structural dominance of some worldviews in environmental discourse.

Additionally, while intersectionality has often centred on gender, race, and class, its application must also grapple with settler colonialism and Indigenous dispossession. This entails expanding intersectional analyses to account for Indigenous epistemologies, governance systems, and rights to land, as well as recognising how colonial and new-colonial systems intersect with patriarchy and capitalism to shape enduring forms of marginalisation. An intersectional approach, hence, should be decolonial—supporting reparatory justice, tribal sovereignty, and knowledge authority to Indigenous peoples.

Moreover, intersectionality requires deep contextual engagement—it is inherently situated and cannot offer one-size-fits-all prescriptions. Its strength lies in its capacity to account for complexity, but this can also pose methodological challenges in implementation. How do planners and policymakers translate intersecting systems of oppression into actionable policies without flattening or instrumentalising them? How can local governments operationalise such complexity while working within constrained institutional structures? These tensions point to the need for participatory, reflexive, and iterative approaches that embrace learning rather than definitive blueprints. A theoretically grounded and critically engaged use of intersectionality should thus maintain its transformative intent while remaining open to critique, self-reflection, and refinement.

Conclusion

As societies confront the compounding crises of environmental degradation, climate emergency and socio-economic inequality, intersectionality offers a valuable lens through which to critically examine how these crises are shaped by and reinforce existing social hierarchies. Currently, strategies for addressing environmental and climate disruptions often remain fragmented, disconnected from transformative priorities, and insufficiently attuned to intersecting inequalities. Rather than proposing a singular agenda, this chapter has explored how intersectional justice can inform both the analysis and practice of socioenvironmental transformation by illuminating pathways toward more equitable and inclusive approaches.

We proposed a fivefold framework for operationalising intersectional thinking in just transformations, encompassing broader environment-society relations:

1. **Challenging systemic gender and race-based inequities** through efforts that address the historical and structural systems that perpetuate racial and gender inequalities, transforming oppressive frameworks that devalue marginalised communities.
2. **Redressing differential vulnerabilities** by identifying and addressing the intersecting impacts of environmental and climate risks, socio-economic crises, and inequitable governance structures, particularly in underserved and marginalised populations.
3. **Valuing care ethics** by recognising care as an essential form of labour, establishing governance systems that distribute care work equitably, and ensuring it is both given and received under fair conditions.
4. **Centring place-based approaches** through planning and policymaking that prioritise local knowledge, decolonise governance processes, and amplify traditionally silenced voices, creating context-specific solutions that reflect diverse experiences.
5. **Fostering cross-identity activism** through civic movements that transcend silos to create inclusive, community-driven resilience efforts that empower marginalised groups to shape strategies and outcomes.

The framework underscores the importance of accounting for power dynamics and systemic inequalities to ensure that all residents, not just those with privilege and resources, can lead secure, dignified, and fulfilling lives in the face of climate, environmental, and societal change.

By adopting intersectional thinking, fragmented strategies and policies can coalesce into integrated approaches that tackle the root causes of vulnerability and inequality. While implementing intersectional approaches will undoubtedly involve negotiation and adaptation, it is an essential step toward dismantling oppressive systems and building more equitable, resilient, and inclusive cities. This framework offers a pathway for scholars, practitioners, and policymakers to operationalise this vision, paving the way for transformative governance and sustainable development.

For Discussion

Intersectionality challenges the idea that focusing on one form of inequality (e.g., gender or race) is sufficient. For instance, anti-racist housing policies might improve access for racialised groups broadly, but fail to account for gendered or familial discrimination—such as the specific exclusion faced by single mothers of colour. In some cases, these policies may even reinforce privilege within a marginalised group by benefiting those closest to dominant norms (e.g., men, higher-income individuals, citizens). Can you think of an example where a policy intended to promote equity ended up benefiting some groups more than others? How can policymakers avoid reinforcing inequalities at the intersections of different systems of oppression?

Intersectionality is not only about marginalization—it also involves analysing the systems that maintain privilege. In climate debates, certain demographics (e.g., affluent, white, cisgender men) are overrepresented in decision-making spaces and among climate sceptics. Understanding how identity, privilege, and power shape attitudes toward climate policy is essential to designing inclusive responses. How can intersectionality help us understand who resists climate action and why? Should climate activism engage more directly with groups that hold systemic power and privilege? How?

References

Adger, W. N., Butler, C., & Walker-Springett, K. (2017). Moral reasoning in adaptation to climate change. *Environmental Politics, 26*(3), 371–390. https://doi.org/10.1080/09644016.2017.1287624

Amorim-Maia, A. T., Anguelovski, I., Chu, E., & Connolly, J. (2022). Intersectional climate justice: A conceptual pathway for bridging adaptation planning, transformative action, and social equity. *Urban Climate, 41*, Article 101053. https://doi.org/10.1016/j.uclim.2021.101053

Amorim-Maia, A. T., Anguelovski, I., Chu, E., & Connolly, J. (2024). Governing intersectional climate justice: Tactics and lessons from Barcelona. *Environmental Policy and Governance*. https://doi.org/10.1002/eet.2075

Amorim-Maia, A. T., Anguelovski, I., Connolly, J., & Chu, E. (2023). Seeking refuge? The potential of urban climate shelters to address intersecting vulnerabilities. *Landscape and Urban Planning, 238*, Article 104836. https://doi.org/10.1016/j.landurbplan.2023.104836

Anguelovski, I., Brand, A. L., Connolly, J. J. T., Corbera, E., Kotsila, P., Steil, J., Garcia-Lamarca, M., Triguero-Mas, M., Cole, H., Baró, F., Langemeyer, J., del Pulgar, C. P., Shokry, G., Sekulova, F., & Argüelles Ramos, L. (2020). Expanding the boundaries of justice in urban greening scholarship: Toward an emancipatory, antisubordination, intersectional, and relational approach. *Annals of the American Association of Geographers, 1–27,*. https://doi.org/10.1080/24694452.2020.1740579

Anguelovski, I., Connolly, J., Pearsall, H., Shokry, G., Checker, M., Maantay, J., Gould, K., Lewis, T., Maroko, A., & Roberts, J. T. (2019). Why green "climate gentrification" threatens poor and vulnerable populations. *Proceedings of the National Academy of Sciences, 116*(52), 26139–26143. https://doi.org/10.1073/pnas.1920490117

BCNUEJ. (2021). *Policy and planning tools for urban green justice*. BCNUEJ.

Berrouet, L. M., Machado, J., & Villegas-Palacio, C. (2017). Vulnerability of socio-ecological systems: A conceptual framework. *Ecological Indicators*. https://doi.org/10.1016/j.ecolind.2017.07.051(Note:missingvolume/issue/pages.)

Blaikie, P., Cannon, T., Davis, I., & Wisner, B. (1994). *At risk: Natural hazards, people's vulnerability and disasters*. Routledge. https://www.preventionweb.net/publications/view/670

Bulkeley, H., Castán Broto, V., & Edwards, G. (2014). *An urban politics of climate change: Experimentation and the governing of socio-technical transitions*. Routledge.

Cannon, C. E. B., & Chu, E. K. (2021). Gender, sexuality, and feminist critiques in energy research: A review and call for transversal thinking. *Energy Research & Social Science, 75*, Article 102005. https://doi.org/10.1016/j.erss.2021.102005

Caprotti, F., & Cowley, R. (2017). Interrogating urban experiments. *Urban Geography, 38*(9), 1441–1450. https://doi.org/10.1080/02723638.2016.1265870

Cardona, O. D., Van Aalst, M. K., Birkmann, J., Fordham, M., McGregor, G., Rosa, P., Pulwarty, R. S., Schipper, E. L. F., Sinh, B. T., Décamps, H., Keim, M., Davis, I., Ebi, K. L., Lavell, A., Mechler, R., Murray, V., Pelling, M., Pohl, J., Smith, A. O., & Thomalla, F. (2012). Determinants of risk: Exposure and vulnerability. In C. B. Field, V. Barros, T. F. Stocker, D. Qin, D. J. Dokken, K. L. Ebi, … P. M. Midgley (Eds.), *Managing the risks of extreme events and disasters to advance climate change adaptation: Special report of the Intergovernmental Panel on Climate Change* (pp. 65–108). Cambridge University Press. https://doi.org/10.1017/CBO9781139177245.005

Cho, S., Crenshaw, K. W., & McCall, L. (2013). Toward a field of intersectionality studies: Theory, applications, and praxis. *Signs, 38*(4), 785–810. https://doi.org/10.1086/669608

Chu, E., Anguelovski, I., & Roberts, D. (2017). Climate adaptation as strategic urbanism: Assessing opportunities and uncertainties for equity and inclusive development in cities. *Cities, 60*, 378–387. https://doi.org/10.1016/j.cities.2016.10.016

Chu, E., Brown, A., Michael, K., Du, J., Lwasa, S., & Mahendra, A. (2019). *Unlocking the potential for transformative climate adaptation in cities*. Global Center on Adaptation. Available here: https://gca.org/reports/unlocking-the-potential-for-transformative-climate-adaptation-in-cities/

Collins, P. H., & Bilge, S. (2020). *Intersectionality* (2nd ed.). Polity.

Connolly, J. J. (2018). From systems thinking to systemic action: Social vulnerability and the institutional challenge of urban resilience. *City & Community, 17*(1), 8–11. https://doi.org/10.1111/cico.12282

Cornwall, A. (2002). Locating citizen participation. *IDS Bulletin.* https://doi.org/10.1111/j.1759-5436.2002.tb00016.x(Note:missingvolume/issue/pages.)

Crenshaw, K. (1989). Demarginalizing the intersection of race and sex: A Black feminist critique of antidiscrimination doctrine, feminist theory and antiracist politics. *University of Chicago Legal Forum, 1989*, 139–167.

Crenshaw, K. (1991). Mapping the margins: Intersectionality, identity politics, and violence against women of color. *Stanford Law Review, 43*(6), 1241–1299. https://doi.org/10.2307/1229039

Davis, K. (2008). Intersectionality as buzzword. *Feminist Theory, 9*(1), 67–85. https://doi.org/10.1177/1464700108086364

Di Chiro, G. (2008). Living environmentalisms: Coalition politics, social reproduction, and environmental justice. *Environmental Politics, 17*(2), 276–298. https://doi.org/10.1080/09644010801936230

Djoudi, H., Locatelli, B., Vaast, C., Asher, K., Brockhaus, M., & Basnett Sijapati, B. (2016). Beyond dichotomies: Gender and intersecting inequalities in climate change studies. *Ambio, 45*(Suppl. 3), 248–262. https://doi.org/10.1007/s13280-016-0825-2

Ducre, K. A. (2018). The Black feminist spatial imagination and an intersectional environmental justice. *Environmental Sociology, 4*(1), 22–35. https://doi.org/10.1080/23251042.2018.1426089

Ehrenreich, B., & Hochschild, A. R. (Eds.). (2004). *Global woman: Nannies, maids, and sex workers in the new economy*. Metropolitan Books.

Enarson, E. (2013). Two solitudes, many bridges, big tent: Women's leadership in climate and disaster risk reduction. In M. Alston & K. Whittenbury (Eds.), *Research, action and policy: Addressing the gendered impacts of climate change* (pp. 63–74). Springer. https://doi.org/10.1007/978-94-007-5518-5_5

Escobar, A. (2001). Culture sits in places: Reflections on globalism and subaltern strategies of localization. *Political Geography, 20*(2), 139–174. https://doi.org/10.1016/S0962-6298(00)00064-0

Federici, S. (2004). *Caliban and the witch.* Autonomedia.

Fisher, B., & Tronto, J. (1990). Toward a feminist theory of caring. In E. K. Abel & M. K. Nelson (Eds.), *Circles of care: Work and identity in women's lives* (pp. 35–62). SUNY Press.

Glenn, E. N. (2012). *Forced to care: Coercion and caregiving in America.* Harvard University Press.

Hankivsky, O. (2014). Rethinking care ethics: On the promise and potential of an intersectional analysis. *American Political Science Review, 108*(2), 252–264. https://doi.org/10.1017/S0003055414000094

Haraway, D. (1988). Situated knowledges: The science question in feminism and the privilege of partial perspective. *Feminist Studies, 14*(3), 575–599. https://doi.org/10.2307/3178066

Held, V. (2006). The ethics of care: Personal, political, and global. *Oxford University Press*. https://doi.org/10.1093/0195180992.001.0001

Johnson, O. W., Han, J.Y.-C., Knight, A.-L., Mortensen, S., Aung, M. T., Boyland, M., & Resurrección, B. P. (2020). Intersectionality and energy transitions: A review of gender, social equity and low-carbon energy. *Energy Research & Social Science, 70*, Article 101774. https://doi.org/10.1016/j.erss.2020.101774

Kabisch, N., van den Bosch, M., & Lafortezza, R. (2017). The health benefits of nature-based solutions to urbanization challenges for children and the elderly: A systematic review. *Environmental Research, 159*, 362–373. https://doi.org/10.1016/j.envres.2017.08.004

Kaijser, A., & Kronsell, A. (2014). Climate change through the lens of intersectionality. *Environmental Politics, 23*(3), 417–433. https://doi.org/10.1080/09644016.2013.835203

Kuran, C. H. A., Morsut, C., Kruke, B. I., Krüger, M., Segnestam, L., Orru, K., Nævestad, T. O., Airola, M., Keränen, J., Gabel, F., Hansson, S., & Torpan, S. (2020). Vulnerability and vulnerable groups from an intersectionality perspective. *International Journal of Disaster Risk Reduction, 50*, Article 101826. https://doi.org/10.1016/j.ijdrr.2020.101826

Lawson, V. (2007). Geographies of care and responsibility. *Annals of the Association of American Geographers, 97*(1), 1–11. https://doi.org/10.1111/j.1467-8306.2007.00520.x

Malin, S. A., Ryder, S., & Lyra, M. G. (2019). Environmental justice and natural resource extraction: Intersections of power, equity and access. *Environmental Sociology, 5*(2), 109–116. https://doi.org/10.1080/23251042.2019.1608420

Malin, S. A., & Ryder, S. S. (2018). Developing deeply intersectional environmental justice scholarship. *Environmental Sociology, 4*(1), 1–7. https://doi.org/10.1080/23251042.2018.1446711

McCall, L. (2005). The complexity of intersectionality. *Signs, 30*(3), 1771–1800. https://doi.org/10.1086/426800

McKittrick, K. (2011). On plantations, prisons, and a black sense of place. *Social & Cultural Geography, 12*(8), 947–963. https://doi.org/10.1080/14649365.2011.624280

Olazabal, M., Chu, E., Castán Broto, V., & Patterson, J. (2021). Subaltern forms of knowledge are required to boost local adaptation. *One Earth, 4*(6), 828–838. https://doi.org/10.1016/j.oneear.2021.05.006

Pelling, M., Chow, W. T. L., Chu, E., Dawson, R., Dodman, D., Fraser, A., Hayward, B., Khirfan, L., McPhearson, T., Prakash, A., & Ziervogel, G. (2021). A climate resilience research renewal agenda: Learning lessons from the COVID-19 pandemic for urban climate resilience. *Climate and Development, 13*(10), 900–908. https://doi.org/10.1080/17565529.2021.1956411

Phoenix, A., & Pattynama, P. (2006). *Intersectionality. European Journal of Women's Studies, 13*(3), 187–192. https://doi.org/10.1177/1350506806065751

Pulido, L. (2017). Geographies of race and ethnicity II. *Progress in Human Geography, 41*(4), 524–533. https://doi.org/10.1177/0309132516646495

Raghuram, P. (2016). Locating care ethics beyond the global north. *ACME: An International Journal for Critical Geographies, 15*(3), 511–513.
Ravera, F., Martín-López, B., Pascual, U., & Drucker, A. (2016). The diversity of gendered adaptation strategies to climate change of Indian farmers: A feminist intersectional approach. *Ambio, 45*(3), 335–351. https://doi.org/10.1007/s13280-016-0833-2
Resurrección, B. P. (2017). Gender and environment from "women, environment and development" to feminist political ecology. In S. MacGregor (Ed.), *Routledge handbook of gender and environment.* Routledge.
Robin, E., & Castán Broto, V. (2020). Towards a postcolonial perspective on climate urbanism. *International Journal of Urban and Regional Research, 45*(5), 869–878. https://doi.org/10.1111/1468-2427.12981
Robinson, C. J. (2000). *Black Marxism: The making of the Black radical tradition.* University of North Carolina Press.
Ruiz-Mallén, I., Baró, F., Satorras, M., Atun, F., Blanc, N., Bortolamiol, S., Casas, L., Clauzel, C., Gallez, E., Grand-Meyer, E., Gras-Velázquez, À., Ivanova, I., Nawrot, T., Presser, P., Reckien, D., & Sekulova, F. (2023). Nature-based solutions for climate adaptation in school environments: An interdisciplinary assessment framework. In Z. Allam (Ed.), *Sustainable urban transitions: Research, policy and practice* (pp. 87–105). Springer Nature. https://doi.org/10.1007/978-981-99-2695-4_6
Sultana, F. (2014). Gendering climate change: Geographical insights. *The Professional Geographer, 66*(3), 372–381. https://doi.org/10.1080/00330124.2013.821730
Sultana, F. (2020). Embodied intersectionalities of urban citizenship: Water, infrastructure, and gender in the Global South. *Annals of the American Association of Geographers, 110*(5), 1407–1424. https://doi.org/10.1080/24694452.2020.1715193
Terry, G. (2009). No climate justice without gender justice: An overview of the issues. *Gender & Development, 17*(1), 5–18. https://doi.org/10.1080/13552070802696839
Thomas, L. (2022). *The intersectional environmentalist*. Voracious.
Thompson-Hall, M. (2016). Land restoration, agriculture, and climate change: Enriching gender programming through strengthening intersectional perspectives. In *Land restoration: Reclaiming landscapes for a sustainable future* (pp. 421–430). Elsevier. https://doi.org/10.1016/B978-0-12-801231-4.00028-8
Timmons Roberts, J., Pellow, D., & Mohai, P. (2018). Environmental justice. In M. Boström & D. J. Davidson (Eds.), *Environment and society: Concepts and challenges* (pp. 233–255). Springer. https://doi.org/10.1007/978-3-319-76415-3_11
Triguero-Mas, M., Anguelovski, I., Cirac-Claveras, J., Connolly, J., Vazquez, A., Urgell-Plaza, F., Cardona-Giralt, N., Sanye-Mengual, E., Alonso, J., & Cole, H.

(2020). Quality of life benefits of urban rooftop gardening for people with intellectual disabilities or mental health disorders. *Preventing Chronic Disease, 17*, E79. https://doi.org/10.5888/pcd17.200087

Virdee, S. (2019). Racialized capitalism: An account of its contested origins and consolidation. *The Sociological Review, 67*(1), 3–27. https://doi.org/10.1177/0038026118820293

Wilson, J., & Chu, E. (2019). The embodied politics of climate change: Analysing the gendered division of environmental labour in the UK. *Environmental Politics, 29*(6), 1085–1104. https://doi.org/10.1080/09644016.2019.1629170

5

Catastrophe

Victor Marchezini and Juan Saavedra

Introduction

Disasters and catastrophes have been examined by scientists across a range of disciplines, yet the engagement of social scientists—particularly sociologists—in disaster studies remains limited. This imbalance is especially evident when comparing research from developed and developing countries. Although the Research Committee on the Sociology of Disasters (established within the International Sociological Association in the 1980s) and the Network of Social Studies on Disaster Prevention in Latin America (La Red, founded in the 1990s) have advanced the field, the participation of scholars from the Global South continues to be scarce. This chapter offers perspectives on the sociology of disasters from South America. While sociology in the region has made significant contributions to contemporary social theory (e.g., dependency theories, decolonization), its engagement with disaster research remains sparse. Drawing on a review of literature from Brazil and Chile, we examine four main aspects: (i) theoretical and methodological approaches, (ii) research subjects, (iii) types of disasters analyzed, and (iv) research topics and

V. Marchezini (✉)
National Early Warning and Monitoring Centre of Natural Hazards (Cemaden), São José Dos Campos, Brazil
e-mail: victor.marchezini@gmail.com; victor.marchezini@cemaden.gov.br

J. Saavedra
Department of Social Sciences, University of Bío-Bío, Concepción, Chile
e-mail: juan.saavedra@ubiobio.cl

D. J. Davidson et al. (eds.), *Environment and Society*,
https://doi.org/10.1007/978-3-032-27059-7_5

disaster phases studied. The study concludes by proposing the creation of a task force to strengthen both disciplinary and interdisciplinary participation of South American sociologists in disaster studies, emphasizing the need for a reparative approach to disaster risk reduction.

One of the earliest sociologists to study disasters was Samuel Henry Prince (1920), who analyzed the social changes that emerged during the recovery process following a devastating ship explosion in Halifax, Canada. The Canadian sociologist defended his dissertation at Columbia University in 1920, offering valuable insights into the relationship between disasters and processes of social transformation. About twenty years later, the Russian sociologist Piririm Sorokin (1942)—then at Harvard University—was studying the historical links between calamities and large-scale social and cultural change. In *Man and Society in Calamity* (1942), Sorokin explored the diversity of human responses to crises such as famine, war, and epidemics, critiquing mechanistic theories that sought to generalize human behavior during disasters.

The "Sociology of Disasters" began to consolidate as a distinct field after the 1950s, particularly in the United States, which promoted a series of studies to analyze how people and organizations prepare for, respond to, and recover from emergencies and disasters (Lindell, 2011; Peek et al., 2021). One of the first primary debates centered on defining what constitutes a disaster. Fritz (1961), for example, described a disaster as an "event" concentrated in time and space in which a society experiences severe danger and losses. These foundational definitions continued to evolve in subsequent edited volumes (Quarantelli, 1998; Perry & Quarantelli, 2005), primarily shaped by scholars from the Global North. A recent overview of the evolution of U.S. disaster sociology is provided by Peek et al. (2021). For those interested in maps and numbers, the web platform of the Social Science Extreme Events Research (SSEER) network (Peek et al., 2024) lists 1,775 social science researchers, mostly based in the U.S. Fewer than 50 are in Latin America and the Caribbean (LAC), and primarily in Chile and Brazil.

Some of the earliest disaster studies emerged after major earthquakes in the 1970s in LAC. Anthropologist Anthony Oliver-Smith, for instance, conducted his doctoral research following the Yungay earthquake in Peru (Faas & Barrios, 2015; Oliver-Smith, 1994, 2012). The occurrence of several disasters in the 1970s and 1980s led social scientists from the United States and Europe to carry out fieldwork across Latin America, often publishing their results in English. In response, Latin American scholars founded the Network of Social Studies for Disaster Prevention (La Red) in 1992 during the International Decade for Disaster Reduction (Lavell, 2005; Maskrey,

1993). La Red brought together researchers and practitioners conducting social analyses of disasters, as summarized in Lavell's (2005) literature review for the Latin American Faculty of Social Sciences (FLACSO). Despite this initiative of La Red, there are few initiatives that seek to convene South American sociologists studying disasters—the last effort was the Research Committee (RC 39), a virtual session at the International Sociological Association (ISA) Forum.

The evolution of disaster studies—as both a field of knowledge and a space of power—has been masterfully examined in *The Invention of Disaster* (Gaillard, 2022). At least three main lines of discussion were identified. The first concerns whether disasters should be described as "natural," a label still used by some scholars but rejected by others who argue that disasters are socially produced. The second concerns the classification of disaster phases—prevention, mitigation, preparedness, response, and recovery. The third asks whether disasters should be understood as discrete events (focusing on emergencies) or as ongoing processes (emphasizing structural root causes). These debates remain foundational. Yet, the escalating impacts of climate change demand renewed attention not only to disasters, but also to catastrophes.

In the aftermath of 2005's Hurricane Katrina, Quarantelli (2006) distinguished between disaster and catastrophe. He stated that catastrophes occur when public services are damaged or destroyed to such an extent that institutions themselves are incapacitated (Oda et al., 2025). Similarly, recent analyses inspired by the COVID-19 syndemic further develop this distinction. Turner (2023), for example, suggests that disasters involve resilience and recovery, whereas catastrophes signify the collapse of institutions and ways of life.

Although these distinctions of magnitude and impacts are valuable, we consider that insights from the sustainability transformations debate (Boström & Lidskog, 2024) can help prevent disasters from escalating into catastrophes. Disaster scholarship has made several contributions to understanding causes and distributions of disasters, as well as barriers to disaster risk reduction (DRR) (Alcántara-Ayala et al., 2023; Wisner et al., 2025). However, discussions on transformative change and disasters are still limited (Impeariale & Vanclay, 2024; Thomalla et al., 2018), particularly those addressing reparative approaches (Eriksen et al., 2025; Hamdanieh et al., 2024; Lampis et al., 2022). Eriksen et al. (2025, p. 1), for instance, call for "reparative disaster governance and management," emphasizing that "a reparative approach seeks to account for the wider historical and systemic conditions that organize and structure the ways disasters unfold, the consequences they bear, and their uneven effects across different people and places."

This chapter draws from a literature review on the sociology of disasters in Brazil and Chile, to identify insights for developing a reparative disaster risk reduction approach (R-DRR). In continuation of the efforts by Gaillard (2022) and Matthewman and Lambert (2024) to foreground Global South perspectives, we focus on three key dimensions: (i) types of disasters analyzed; (ii) theoretical and methodological approaches; and (iii) research topics and disaster phases explored. The chapter concludes with a proposal to establish a task force that brings together South American sociologists and anthropologists, encouraging collaboration with marginalized communities to co-create an action research agenda on reparative disaster risk reduction.

Social Studies of Disasters in Brazil

Brazil has 5,570 municipalities distributed across five regions (South, Southeast, Central-West, North, and Northeast) and six main biomes—the Amazon Forest (Tropical Forest), Savannah (Cerrado), Caatinga (semiarid), the Atlantic Forest, Pantanal (tropical wetland), and Pampas (grasslands). These municipalities are exposed to a variety of socio-natural hazards, including droughts, heatwaves, riverine floods, flash floods, coastal floods, wildfires, tornadoes, thunderstorms, landslides, debris flow, erosion, and other mass movements. Changes in temperature and rainfall patterns may intensify these hazards (Marengo et al., 2022; Vasconcelos Junior et al., 2024). Vulnerabilities to these hazards differ across Brazil (Hummell et al., 2016), both within regions—for instance, the recurrent droughts in the Brazilian Northeast (Vieira et al., 2020)—and within cities. A national survey conducted across 821 municipalities estimated that roughly eight million people live in landslide- and flood-prone areas classified as high or very high risk (Assis Dias et al., 2018). This study, developed by the Brazilian Institute for Geography and Statistics (IBGE) in partnership with the National Center for Early Warning and Monitoring of Natural Hazards (Cemaden), is based on census tract data provided by IBGE. Although the IBGE census has data about race, this variable was not incorporated into the survey on people living in flood- and landslide-prone areas in Brazil. This omission—what Gantus-Oliveira (2024) calls a *"race-blind"* approach to disasters—represents a significant knowledge gap for developing a reparative disaster risk reduction (R-DRR) framework. The unequal distribution of risk can also be observed at the neighborhood level—particularly in favelas (slums). These inequalities can be traced to both the root causes of vulnerability (such as income inequality and unequal land distribution) and to dynamic pressures (including urban sprawl, real estate speculation, and land grabbing) that force marginalized populations to inhabit unsafe areas. Many of these communities are located on steep slopes and lack access to basic sanitation (Marchezini & Wisner, 2017; Wisner et al., 2025).

Types of Disasters Analyzed

Most disasters and catastrophes in Brazil are associated with hydrometeorological hazards, such as droughts, floods, and landslides. Between 1991 and 2023, 4,958 of the country's 5,570 municipalities (approximately 89%) were affected by water-related disasters, impacting a total of 232 million people. Some of these disasters occurred simultaneously across 16 of Brazil's 26 states, requiring coordinated response and recovery efforts from national and subnational governments (Marchezini et al., 2025; Oda et al., 2025). For instance, between January and March 2004, floods affected 1,224 municipalities, displaced 350,000 people, destroying 114,000 houses across 16 states (Brasil, 2004).

Flood- and landslide-related disasters have also caused severe damage on a regional scale. In May 2024, approximately 95% of the municipalities in Rio Grande do Sul were affected by widespread flooding. Similarly, in 2008, major floods and landslides struck the Itajaí Valley in Santa Catarina State, resulting in over 250 fatalities. Table 5.1 summarizes several of these water-related disasters and catastrophes.

Although these numbers and statistics are continuously updated and publicly available on the website of the National Secretariat of Civil Defense and Protection (Sedec), the lack of data on gender, race and age among people affected by disasters presents a barrier to generating knowledge for reparative DRR. Some studies such as Couto (2023) have addressed this gap by using alternative government databases to estimate the number of women who have died in water-related disasters in Brazil.

Some of these disasters and catastrophes have catalyzed political shifts in Brazil's disaster risk management (DRM) policy. In January 2011, when more than 1,000 people died during floods and landslides in the Mountain Region of Rio de Janeiro State, newly elected President Dilma Rousseff called on several ministries to develop the National Plan on Risk Management and Disaster Response (PNGRD). As a direct result of this catastrophe,

Table 5.1 Water-related disaster and catastrophes reports in Brazil, 2000–2024

Year	Hazard	Identification	Most affected location	State or region affected
2004	Climatic	Floods and disasters	1,000 cities	Several states
2004	Climatic	Hurricane	Florianópolis and nearby cities	Santa Catarina
2005	Climatic	Drought	739 municipalities	South Region
2005	Climatic	Drought	several municipalities	Amazon region
2008	Climatic	Floods and landslides	37 cities of Itajai Valley	Santa Catarina
2009	Climatic	Drought	several municipalities	Amazon region
2010	Climatic	Floods	100 cities	Alagoas and Pernambuco
2010	Geological	Landslides	Angra dos Reis and Niterói	Rio de Janeiro
2010	Climatic	Floods	São Luiz do Paraitinga	São Paulo
2011	Climatic and Geological	Floods and landslides	Nova Friburgo, Petrópolis, Teresópolis, São José do Vale do Rio Preto and others	Rio de Janeiro
2012	Climatic	Floods	several municipalities	Acre and Rondônia
2013	Geological	Floods and landslides	several municipalities	Espírito Santo
2013	Climatic	Drought	several municipalities	Northeast
2014	Climatic	Floods	Rio Branco and other cities	Acre
2014	Climatic	Drought	several municipalities	Northeast
2015	Climatic	Drought	several municipalities	Northeast
2016	Climatic	Drought	several municipalities	Northeast
2017	Climatic	Drought	several municipalities	Northeast
2021	Climatic	Floods	several municipalities	Bahia
2022	Climatic	landslides	Petrópolis	Rio de Janeiro

(continued)

Table 5.1 (continued)

Year	Hazard	Identification	Most affected location	State or region affected
2023	Climatic	landslides	São Sebastião	São Paulo
2024	Climatic and Geological	Floods and landslides	95% of municipalities of Rio Grande do Sul	Rio Grande do Sul

Source Prepared by the authors based on data from the Disaster Information System (S2ID)

the National Policy on Civil Defense and Protection (PNPDEC) was enacted the following year (Marchezini et al., 2017).

Earlier events, such as the 2008 disaster in parts of Santa Catarina State and the 2011 catastrophe in Rio de Janeiro, also inspired numerous theses and dissertations examining the causes and impacts of disasters through diverse disciplinary lenses. Rodrigues et al. (2015) stated that 92% (83) of theses and dissertations on disaster-related topics were defended between 2008 and 2013. However, research in sociology, anthropology, and political science accounted for only 10% of the total. As noted by Taddei (2014) and Ribeiro (2020) in anthropology, and by Marchezini (2018) in sociology, the social sciences in Brazil have produced relatively few studies on disasters. The following subsection examines several sociological approaches to disasters in the Brazilian context.

Theoretical, Methodological Approaches to the Sociology of Disasters in Brazil

One of the earliest dissertations in Brazil to study disasters from a sociological perspective was written by Marcos Mattedi (1999). Personally affected by the 1983 flood in the Itajai Valley, Santa Catarina, Mattedi examined the historical process of land use change and public policy that shaped patterns of social exposure and vulnerability to floods, as well as the governmental responses aimed at reducing impacts and preventing new disasters. His research relied on archival analysis of newspaper discourses about floods in the Itajaí Valley during the nineteenth and twentieth centuries.

The first Brazilian research group dedicated to analyzing disasters through sociological lenses (known as NEPED/Ufscar) was founded by Norma Valencio in 2004. Valencio (2012) framed the sociology of disasters as a branch of the sociology of development, conceptualizing disasters as social processes rooted in development models and their consequences: rising inequality, overexploitation of natural resources, land grabbing, real estate speculation, fragile livelihoods, precarious settlements, and poverty. Her approach aligns closely with frameworks advanced by Latin American scholars from La Red scholars (Alcántara-Ayala et al., 2023; Garcia-Acosta, 1997, 2008, 2020; Lavell, 1993) and with the Pressure and Release model (Blaikie et al., 1994; Wisner et al., 2004, 2012, 2025).

In her book *Para Além do "Dia" do Desastre* (*Beyond the "Day" of the Disaster*) (Valencio, 2012), Valencio synthesizes a decade of disaster research in Brazil, employing both quantitative and qualitative methods to analyze prevention, mitigation, preparedness, response, and recovery. Drawing on Pierre Bourdieu's (1991) reflections on power, she examined relations among civil defense agencies, the press, and affected communities to reveal forms of symbolic violence and inequality embedded in disaster management discourse.

At NEPED/Ufscar, the studies conducted by many of Valencio's students have explored social dimensions of shelter administration, including age, gender, disability, and other intersectional factors. Siena (2009) examined gender relations and power dynamics between shelter managers and residents, while Marchezini (2010) drew on Bachelard (1989) and Bourdieu (1991) to analyze how disaster survivors sought to reproduce private, domestic practices within the public spaces of temporary shelters, often set up in schools. Geraldi (2009) identified the particular challenges faced by people with disabilities in these temporary living environments.

Recognizing these vulnerabilities, NEPED/UFSCar later expanded its research focus to encompass both short- and long-term disaster recovery. Revisiting past disaster sites, Valencio et al. (2011) identified several human rights violations among those affected, who had waited years for housing reconstruction. The authors referred to this group as "abandonados nos desastres" (people abandoned in disasters) (Valencio et al., 2011). Other studies have analyzed the roles of community-based organizations, national and local non-governmental organizations (Brenes, Marchezini & Wisner, 2020), and emerging social movements of disaster-affected people (Marchezini, 2019) advocating for justice and reparation in disaster contexts. These findings underscore the need for greater public debate on reparative responses to water-related disasters.

As in other areas of disaster research, social studies on disaster recovery processes remain scarce in Brazil. Marchezini (2013, 2015, 2019) used a

Foucauldian approach to examine power dynamics, discourses, and practices among social agents in the aftermath of the 2010 flood in Sao Luiz do Paraitinga, Brazil. He observed that, during the emergency phase, public officials and the media focused primarily on heroic narratives of "saving lives." Yet, weeks later, the disaster's human and material losses were deprioritized, leaving affected populations to fend for themselves amid a logic of "letting die"– a form of necropolitics (Mbembe, 2019) that exposes deeper structures of abandonment. Marchezini termed this the "biopolitics of disaster," referring to the mechanisms and dispositifs within the social body that classify people, collect data, and manage biological life in ways that serve political and economic interests.

These mechanisms of biopolitical control can hinder social movements and marginalized groups from mobilizing for transformative change, as they often rely on state-controlled data that obscures inequality. One example occurred during the Covid-19 pandemic, when the lack of data on deaths in Brazilian favelas and the absence of adequate social protection policies for poor populations (Muniz et al., 2021; Marchezini et al., 2026) exacerbated vulnerability. Another case is the catastrophe of the Rio Doce river basin following the Mariana dam collapse on November 5, 2015. The mining companies Samarco, Vale, and BHP Billiton established the Renova Foundation, a nonprofit organization tasked with managing recovery efforts. However, much of the data on losses and damages was controlled by the foundation, framing the catastrophe in ways that simplified its complexity and downplayed the need for holistic reparation (*see Box* 1).

Box 1: Rio Doce watershed catastrophe and the need for a reparative, holistic approach

On November 5, 2015, approximately 43 million m^3 of mining tailings—about 80% of the total contained volume—were released during the Mariana dam collapse. The resulting 10-meter-high mud waves caused irreversible environmental damage to hundreds of watercourses within the Doce River basin and its associated ecosystems (Carmo et al., 2017). The disaster led to the mass death of fish from 21 different species (IBAMA, 2015) and the loss of several large mammals (Carmo et al., 2017).

The tailings directly damaged 863.7 ha of Permanent Preservation Areas associated with watercourses. Additionally, 294 small creeks and forty downstream cities were damaged, leaving hundreds of thousands of people (including indigenous communities) without access to clean water. At least 806 buildings were damaged, and 218 completely destroyed, including homes, public buildings, commercial properties, centennial churches and ancient farms across 10 districts in five municipalities. Cultural heritage sites also suffered severe damage. Including two archeological sites, six places of historical and cultural interest, and more than 2,000 pieces of sacred material heritage (Fernandes et al., 2016).

Scientific studies, including those from sociology and anthropology studies (Branco, 2023; Losekann, 2017; Losekann & Milanez, 2023; Zhouri et al., 2017), have documented the wide range of social and environmental impacts, as well as the ongoing struggles of affected communities seeking reparation. As with many other disasters and catastrophes in Brazil, governmental discourses and practices following the Mariana dam catastrophe have often failed to incorporate a reparative and transformative approach that challenges the existing status quo.

Studies examining disaster recovery processes have also identified forms of disaster capitalism, referring to the exploitation of disasters as opportunities for economic, social, or political gain, often through measures such as privatization, deregulation, and urban redevelopment (Klein, 2007; Sandoval et al., 2020; Schuller, 2008). Only a few studies have analyzed expressions of disaster capitalism in Brazil (Marchezini, 2015; Rojas, 2020).

Another key topic within the Brazilian sociology of disasters is disaster governance. Pismel et al. (2023) analyzed wildfire governance in the tri-national frontier region of Madre de Dios (Peru), Acre (Brazil), and Pando (Bolivia), and Bravo (2022) studied the governance of landslide warning systems and its limitations at the national level. However, there remains a lack of research examining governance across other hazards and phases of disaster risk management (DRM), as well as studies that bridge reparative disaster governance and adaptive governance approaches.

Finally, sociologists of disasters face several barriers. Marchezini (2024) notes how some sociologists question the legitimacy of studying disasters as a sociological topic. Similar barriers arise when sociologists engage with researchers from other disciplines in disaster studies. Power imbalances between fields have been documented in interdisciplinary projects (Peek & Guikema, 2021; Peek et al., 2020). For example, Marchezini (2020) used institutional ethnography to describe his experience at a warning agency, where colleagues frequently asked him, *"What is a sociologist doing here?"* Such encounters highlight how disciplinary hierarchies can limit collaboration. In the context of reparative disaster risk reduction, these imbalances—between both disciplines and social positions—must be acknowledged and addressed.

Social Studies of Disasters in Chile

Chile has a distinctive geographic shape extending over 4,270 km in length, but with a width that does not exceed 90 km. It is a unitary sovereign state, administratively divided into 16 regions. According to the most recent population census (2024), the country has 18,480,432 inhabitants. The capital and principal city is Santiago, where approximately 41% of the country's population resides in the urban area. The country's geography introduces several key challenges for DRM. First, Chile exhibits significant climatic diversity across continental territory, according to the Köppen-Geiger classification, and second, it lies at the convergence of the Nazca and South American tectonic plates. This geographical context generates a wide range of natural hazards that interact with the country's social, economic, and institutional vulnerabilities. Historically, these interactions have produced catastrophes of considerable significance for both the scientific community and international humanitarian organizations.

Major earthquakes have been recorded in 1730, 1868, and 1960, while extreme precipitation events associated with flooding occurred in 1965, 1982, and 1991. Severe droughts were also registered in 1924 and 1968–1969. Additionally, the country faces ongoing volcanic hazards, with more than 90 active volcanoes distributed throughout its territory.

Types of Disasters Analyzed

We identified thirty-four significant disasters in Chile between 2000 and 2024. Most were related to climate hazards, which have become increasingly frequent since the 2010s. According to official records, the 2010 earthquake and tsunami caused the highest number of fatalities during this period, with 527 deaths. More recently, severe wildfires in 2023 and 2024 resulted in 164 fatalities. This list does not include the Covid-19 syndemic (2020–2023). Table 5.2 summarizes these events by year and affected territory.

At the subnational level, the Los Lagos, Biobío, and Valparaíso regions appear as the most affected in this list, having experienced various types of disasters throughout the twenty-first century. Geologically-induced events are concentrated in regions such as Maule, Biobío, Los Lagos, and Tarapacá, while extreme climatic events have primarily affected the Valparaíso and Biobío regions. Both zones are among the most densely populated in the country. In the latter cases, the high incidence of disasters may be related to rising temperatures, decreasing rainfall, and the management of human

Table 5.2 Disaster reports in Chile, 2000–2024

Year	Hazard	Disaster type	Most affected location	Region in Chile
2002	Climatic	Buill flood	Buill, Chaitén	Los Lagos
2005	Geological	Tarapacá earthquake	Iquique	Tarapacá
2006	Climatic	Biobío flood	Los Ángeles	Biobío
2006	Climatic	Landslide	Chiguayante	Biobío
2007	Geological	Aysén earthquake	Puerto Aysén	Aysén
2007	Geological	Tocopilla earthquake	Tocopilla, Antofagasta	Antofagasta
2008	Geological	Chaitén eruption	Chaitén	Los Lagos
2010	Climatic	Drought	Most of Chile	Country
2010	Geological	27F earthquake	Constitución, Talcahuano	Maule-Biobío
2011	Geological	Puyehue eruption	Futrono, Rio Bueno	Los Lagos
2012	Climatic	Torres del Paine fire	Torres del Paine	Magallanes
2012	Climatic	Río Las Minas flood	Punta Arenas	Magallanes
2014	Geological	Arica-Iquique earthquake	Iquique	Parinacota-Tarapacá
2014	Climatic	Urban fire	Valparaíso	Valparaíso
2015	Climatic	Antofagasta flood	Antofagasta	Antofagasta
2015	Climatic	Atacama flood	Chañaral-Copiapó	Atacama
2015	Geological	Coquimbo earthquake	Tongoy-Canela	Coquimbo
2015	Geological	Calbuco eruption	Ensenada	Los Lagos
2015	Geological	Villarrica eruption	Pucón-Coñaripe	Araucanía
2015	Climatic	Cajón del Maipo landslide	San José de Maipo	Metropolitana
2016	Geological	Quellón earthquake	Quellón	Los Lagos
2017	Climatic	Atacama flood	Chañaral	Antofagasta
2017	Climatic	Urban fire	Valparaíso	Valparaíso
2017	Climatic	Maule-Biobío fire	Santa Juana	Biobío

(continued)

Table 5.2 (continued)

Year	Hazard	Disaster type	Most affected location	Region in Chile
2017	Climatic	Santa Lucia landslide	Chaitén	Los Lagos
2019	Climatic	Tarapacá landslide	Camiña	Tarapacá
2019	Climatic	Limache forest fire	Limache	Valparaíso
2019	Climatic	Urban fire	Valparaíso	Valparaíso
2019	Climatic	Biobío tornadoes	Talcahuano	Biobío
2022	Climatic	Urban fire	Viña del Mar	Valparaíso
2023	Climatic	Fire	Center-south	Ñuble-Biobío-Araucanía
2023	Climatic	Flood	Center-south	Biobío Ñuble Maule
2024	Climatic	Fire	Peñuelas-Viña del Mar-Quilpué	Valparaíso
2024	Climatic	Heavy rainstorms	Licantén	Maule

Source Prepared by the authors, based on information from Chilean State agencies

settlements (Ubilla-Bravo & Rodríguez-Seguel, 2022). Geological disasters, by contrast, tend to occur less frequently but cause greater devastation owing to their lower predictability and wider geographic reach.

Geological-Related Disasters

Among Chile's geological disasters, the most significant was the earthquake of February 27, 2010. The quake's epicenter was located near the coastal town of Cobquecura, on the border between the Maule and Ñuble regions in the south-central part of the country, and reached a magnitude of 8.8Mw. As a subduction zone earthquake, it generated a highly destructive tsunami that struck the central and southern coasts. This event marked a turning point in recent Chilean history, symbolically occurring during the nation's bicentennial year. In the aftermath of the earthquake, signs of the breakdown of the post-dictatorship social pact became evident, foreshadowing the social unrest that emerged in 2019. A second major case was the earthquake of September 2015, which struck north-central Chile with a magnitude of 8. 4Mw.The main affected areas included the coastal cities of Coquimbo and La Serena, as well as the surrounding inland rural towns. Following the earthquake, a destructive tsunami again impacted the Coquimbo coast. The most notable volcanic eruption in recent decades occurred in May 2008 at Chaitén

Volcano, which released gas and ash rising more than 20 km high and forced the complete evacuation of the nearby town. Other major eruptions include those of the Puyehue–Cordón Caulle volcanic complex in 2011 and Calbuco Volcano in 2015.

Climate-Related Disasters

One of Chile's most significant climate-related disasters was the prolonged drought between 2010 and 2023, the longest in the country's recorded history. This drought led to severe water shortages, particularly in the Coquimbo, Valparaíso, and O'Higgins regions, and drastically affected the supply of drinking water in rural areas. It was also linked to devastating wildfires in the heavily forested and forestry-exploited zones of Maule, Ñuble, Biobío, and La Araucanía. Another extreme climate event was the intense rainstorm that struck the Atacama Desert on March 24 and 25, 2015. The resulting floods caused widespread damage, especially in the towns of Copiapó, Diego de Almagro, and Chañaral, where public infrastructure and housing were severely affected.

The most destructive wildfires occurred in 2017 and 2023, impacting the O'Higgins, Maule, and Biobío regions. Extreme heat and strong winds exacerbated these fires, while ongoing drought and rising temperatures have contributed to the increasing frequency of large-scale wildfires across the country. The most affected areas—primarily under intensive forestry exploitation—reflect Chile's capitalist-extractivist economic model. The most notable urban–rural interface fires occurred in Valparaíso in 2014 and Viña del Mar in February 2024, causing 137 fatalities. Other notable climate-related events include the 2006 rainstorms in central-southern Chile, which caused the Biobío River to overflow, flooding areas in Los Ángeles and along the riverbanks of Concepción. This region also experienced the century's most significant tornadoes and waterspouts in May 2019.

Theoretical, Methodological, and Subject Approaches in the Sociology of Disasters in Chile

As noted earlier, the twenty-first century in Chile has seen significant disasters that have drawn the attention of social scientists. This period coincides with the political and cultural consolidation of neoliberalism (Madariaga, 2020; Posner, 2023) established during the last military dictatorship (1973–1990). During this time, the neoliberal regime deepened, prioritizing the market over other aspects of social life. This context is crucial for understanding how the extractivist model contributes to the production of disasters (Bradshaw, 2024).

Romero and Romero (2015) observe that social science studies on disasters in Chile have emerged only gradually. These disciplines have integrated into a transdisciplinary field initially led by engineering and earth sciences. Geography has played a central role in advancing both scientific research and political debate about disasters (e.g., Martínez et al., 2020; Romero-Aravena & Mendes, 2021). Sociology has increasingly engaged with disaster studies within its own field and through collaborations with other disciplines. Notable contributions include those of Tironi (2022, 2023), Rivas (2022), and Azocar et al. (2021), among others. For example, Tironi and Undurraga (2023) emphasize the importance of integrating geology into Chilean sociology to enrich discussions of the Anthropocene. Studies by sociologists such as Azocar et al. (2021) examine the social perceptions of communities vulnerable to climate change. Other fields, including psychology (e.g., Barrales, 2019; Berroeta & Pinto De Carvalho, 2020), anthropology (Osorio-Parraguez, 2016), and social work (e.g., Moreno, 2018, 2019; Sepúlveda, 2024), have also contributed to disaster research.

Like other social scientists, sociologists often draw on shared theoretical frameworks. As in other Latin American countries (e.g., Soares-Guimarães et al., 2025), political ecology has emerged as Chile's most influential theoretical approach to disaster studies. This framework examines the intersections of politics, economy, and environment, focusing on social and spatial justice, critiques of neoliberalism, and the integration of diverse forms of knowledge to understand the relationship between society and nature. Romero et al. (2015) argue that extreme natural events become risks depending on the location and activities of human communities. Bustos et al. (2015) propose an ontological perspective on the society-nature relationship, criticizing the subordination of science to neoliberal logics, while Rivas et al. (2024) suggest that climate change and environmental crises in Chile have increasingly blurred the distinction between natural and human-made threats. Other critical approaches in Chilean disaster studies include discussions on decolonization and social-spatial justice.

Andueza (2023) argues that Chile faces a profound socio-political crisis driven by neoliberalism and climate change, where elites have little incentive to adopt environmentally responsible actions in public disaster response policies (Mena, 2018). Bravo and Fragkou (2019) conducted an important study on water scarcity, focusing on gender, Mapuche cosmovision, and political ecology. Water is a vital element in the ecological and social balance of the Mapuche people, Chile's largest Indigenous group (Ibarra & Riquelme, 2019). The ecopolitical framework provides an important lens for understanding disasters that have occurred in Chile over recent decades.

The deepening of capitalism in countries such as Brazil and Chile has driven political reforms across all spheres of social and institutional life. Since the 1970s, privatization has increased while the role of the state has been reduced, and police repression of social movements has intensified. In Chile, the neoliberal regime has profoundly shaped disaster management (Saavedra et al., 2021). This influence is particularly visible in how market-driven decisions guide responses to catastrophes. Panez (2022) argues that neoliberalism, established in the 1970s, has become rigid and less adaptable to changing contexts. Campos et al. (2021) made a valuable contribution to post-disaster studies by analyzing the aftermath of the 2014 urban fire. They criticized the neoliberal reconstruction model, arguing that it displaced affected populations rather than addressing their needs—a critique rooted in reparative approaches identified in studies on disaster capitalism in Chile (González-Muzzio & Sandoval, 2018) and on the neoliberal biopolitics of reconstruction (Saavedra, 2023).

The complexity of disaster scenarios in Chile calls for a coordinated approach at national and regional levels, aligned with international frameworks for prevention, mitigation, and recovery. As a broad interpretive paradigm, political ecology has shaped social science research on disaster risk governance and aligns with the priorities of the Sendai Framework for Disaster Risk Reduction 2015–2030. The global perspective has also influenced Chile's Law No. 21.364 (2021), which established the National System for Disaster Prevention and Response. This law enables the development of policies, plans, and regulations addressing every phase of the disaster cycle. DRM is understood as a continuous process focused on reducing hazards arising from both natural and human-made threats. It also promotes coordination between private and public actors, encourages citizen participation, and strengthens institutional capacities to manage disasters through comprehensive and sustainable approaches. Other regulations, such as Law No. 21.455 on Climate Change Framework and NCH 433 on seismic building design, further support these efforts.

Social scientists have critically examined the network of plans and regulations. Vergara and Ejsmentewicz (2024) introduce the concept of the "hyper-object" to capture the multidimensional nature of disasters, emphasizing the challenges of implementing governance models. They point to the *dislocation of catastrophe*, as exemplified in the response to the 2010 earthquake. Sandoval (2016, 2020) criticizes Chile's centralized approach to disaster planning and response, arguing that poor coordination between institutional and community actors undermines disaster management efforts. In their study of the 2008 Chaitén volcanic eruption, Sandoval and Voss (2018) found that inadequate planning exacerbated unsafe conditions and deepened mistrust among affected communities.

Research on resilience in Chilean disaster contexts often focuses on the preparedness of local and community actors, institutional response capacity and the material and symbolic resources available in affected areas. Micheletti and Letelier (2016) examined post-earthquake housing reconstruction in the Maule-Biobío region after the 2010 earthquake, showing that social research on resilience spans multiple disciplines. Other studies address challenges in Chile's expanding tourism sector, one of the most important in Latin America, where robust emergency systems are needed to protect visitors and ensure preparedness in tourism-dependent regions (Alcota & Aravena-Reyes, 2020).

Resilience has also been examined in the context of crises such as the COVID-19 pandemic. Fuentes Aguilar et al. (2024) emphasize that community adaptation and collective learning are key to facing crises but note that social cohesion and mutual support declined during the pandemic due to enforced distancing measures.

Gender perspectives are critical in Chile's disaster research. Incorporating a gender lens helps to address power dynamics and strengthen women's resilience over time (Moreno & Shaw, 2018). Intersectional approaches enrich this discussion by revealing how overlapping inequalities shape women's experiences of disasters (Vergara et al., 2022). While traditional disaster sociology has focused on structural vulnerability and resilience, Vergara et al. (2022) refine this perspective by integrating gender and intersectionality, exposing inequalities and challenging the assumed neutrality of conventional categories. These frameworks also promote women's leadership in disaster response, as illustrated by community kitchen initiatives during the Covid-19 syndemic (Inzunza et al., 2023). Moreno and Shaw (2018) also show how the 2010 earthquake heightened women's vulnerabilities but also created opportunities for leadership. Similarly, Acuña et al. (2021) document how women in Santiago settlements led risk management efforts, addressing threats such as the pandemic, fires, and landslides.

The relationship between disasters and indigenous cultural expression in Chile is another essential theme. Sandoval et al. (2024) highlight that indigenous communities often occupy vulnerable territories, making them more susceptible to droughts, floods, and wildfires. The Mapuche, Chile's largest indigenous group, inhabit a vast territory spanning southern Chile and parts of Argentina. Their creation myth, centered on the battle between the serpents *Tren Tren and Kai Kai Vilu*, provides a cosmological explanation for earthquakes and tsunamis (Inostroza-Matus et al., 2019). Social, cultural, and historical dimensions of Indigenous life are therefore central to understanding disaster dynamics in Chile.

Box 2: Indigenous communities in disaster studies

Case study 1

Cadierno-Gutiérrez (2018) examined the experiences of the Mapuche-Lavkenche community in past disasters, such as the 1960 earthquake. The Lavkenches' knowledge of earthquakes and tsunamis is transmitted through oral tradition, lived experience and intergenerational teaching. Myths such as Kai-Kai and Treng-Treng embody cosmological interpretations that guide precautionary measures. Kimün (ancestral knowledge) and piam (historical narratives) are shared by elders, community leaders and intercultural schools. This indigenous knowledge proved vital during the 2010 tsunami, when most of the population evacuated to safe areas. The combination of ancestral wisdom and community communication contributed significantly to disaster preparedness.

Case study 2

In Chile's Atacama Desert, Ramos-Chocobar and Tironi (2023) describe differences between Indigenous and Western understandings of disasters. Lickanantay volcanology is an ancestral system that links volcanoes, humans, climate, water, and agriculture within a geocosmic network of interdependence. As ancestral beings and extended family (kin), volcanoes are central to a cosmology in which the *sol interior* (Inner Sun) and *Pachamama* (Mother Earth) maintain ecological and spiritual balance. This worldview opposes extractivism and redefines volcanic risk as the disruption of relational cycles. Rooted in histories of Chileanization and epistemic colonialism, it supports global efforts to integrate Indigenous knowledge into disaster governance while affirming Lickanantay territorial and epistemic sovereignty.

Case study 3

For the Aymara people, Mondaca and Díaz (2022: 61) note that these communities have developed an "ethnoperception" that includes geographic selectivity, useful for disaster prevention, particularly in climate-related events. The Aymara perceive the climate through a holistic worldview, integrating cultural, ecological, and spiritual dimensions. Key concepts such as Chacha-Warmi (gender complementarity) and Pacha (the living cosmos) shape their understanding of atmospheric phenomena and guide their agricultural decisions. This ancestral knowledge, grounded in sustained empirical observation, draws on indicators such as cloud formations, wind patterns, water sources, flora, and fauna. Unlike technocentric approaches, it constitutes an adaptive system that maintains social and ecological resilience in the Andean highlands, transmitting intergenerational strategies for living in balance with environmental variability.

Conclusions and Recommendations

Currently, there are no comprehensive efforts to map sociological studies on disasters in Brazil and Chile. Establishing such a mapping process—and creating a network of South American sociologists working on disasters—would foster greater exchange in research, teaching, and activism. This chapter sought to contribute to this initiative through a literature review that identifies key sociological studies on disasters. Although the scope was limited to climatic and geological hazards, the review offers a meaningful overview of theoretical and methodological approaches as well as central research topics.

A comparison between Brazil and Chile reveals two comparable lines of thought. The first is a more empirical approach that identifies gaps in research and offers insights from critical poststructural frameworks (e.g., biopolitics). The second is an equally critical approach grounded in political ecology, critiques of neoliberalism, and indigenous worldviews on disasters. These complementary perspectives provide a foundation for a regional sociological approach to disasters in South America.

Mapping existing networks would help strengthen cooperation among social scientists studying diverse dimensions of disaster. Expanding these connections could enhance the social sciences' capacity to address emerging issues, including: (1) critically revisiting prevention, mitigation, and recovery efforts in the context of weakened public institutions shaped by neoliberal ideologies; (2) integrating Indigenous cosmovisions and ethnographic studies; (3) addressing technological gaps among communities, businesses, and governments that constrain disaster management in countries such as Brazil and Chile, recognizing that technology, when integrated into social disaster research, can improve analytical capacity and support the design of context-specific recovery plans; and (4) advancing an action research agenda on reparative disaster risk reduction (R-DRR), aimed at transforming structures of oppression and improving access to resources such as water, land, housing, and energy for indigenous people, and Black and Brown communities (Fitz-Henry & Klein, 2024).

Future research should continue building toward reparative disaster risk reduction by mapping sociologists of disasters in other countries across Latin America and the Global South. Organizing webinars and workshops would be essential for connecting researchers, promoting comparative studies, developing innovative methodologies, and addressing common challenges—such as disciplinary power imbalances, lack of funding, and other structural barriers to collaboration.

Discussion Questions

Consider how determinations of an event as a disaster or catastrophe are subjective processes. Can you think of an extreme event that might be defined as a disaster or catastrophe in one place and time, but might not be defined as such in another?

Among the many case studies of disaster presented in this chapter, can you identify examples of the four types of injustice described in Chapter 2?

How can just transformation be realized in a world of increasing frequency and intensity of climate-attributed disasters? In what ways do such events compromise the collective societal capacity for just transformation? Might the occurrence of disaster or catastrophe enhance the likelihood for just transformation?

Acknowledgements Victor Marchezini acknowledges the São Paulo Research Foundation (Fapesp) for financial support and funding (Grant number #2022/02891-9).

Juan Saavedra acknowledges the ANID Chile for financial support and funding (Grant number Fondecyt #1251794).

References

Acuña, V., Valdivieso, S., & Juzam, L. (2021). Dignificando la gestión de riesgo de desastres: Liderazgos femeninos y estrategias comunitarias en el Campamento Dignidad, Santiago de Chile. *REDER, 5*(2), 91. https://doi.org/10.55467/reder.v5i2.79

Aguilar, A. F., Saravia, R. S., Riffo, I. A., & Acevedo, J. B. (2024). Estrategias de gestión del riesgo en Concepción. Afrontamiento familiar y comunitario en familias vulnerables durante la pandemia del COVID-19. *Revista de Estudios Latinoamericanos sobre Reducción del Riesgo de Desastres REDER, 8*(1), 180–192. https://doi.org/10.55467/reder.v8i1.146

Alcántara-Ayala, I., Gomez, C., Chmutina, K., van Niekerk, D., Raju, E., Marchezini, V., Rom Cadag, J., & Gaillard, J. (2023). *Disaster risk* (1st ed.). Routledge. https://doi.org/10.4324/9781315469614

Alcota, P., & Aravena-Reyes, A. (2020). Desastres, memorias y bienestar de lugar en Dichato. *Chile. Interdisciplinaria, 37*(1), 17–18.

Almeida, L. Q. D., Araujo, A. M. S. D., Welle, T., & Birkmann, J. (2020). DRIB Index 2020: Validating and enhancing disaster risk indicators in Brazil. *International Journal of Disaster Risk Reduction, 42*, Article 101346. https://doi.org/10.1016/j.ijdrr.2019.101346

Andersen, K., Rodríguez-Negrete, L., & Balbontín-Gallo, C. (2021). Las mujeres en la reconstrucción del espacio público post-catástrofe socio-natural en Dichato, Chile (2010–2013): Una aproximación hermenéutica desde el mito de Antígona. *Izquierdas, 49*, 725–742. https://doi.org/10.4067/s0718-50492020000100239

Andueza, L. (2023). Notes on climate, late neoliberalism, and the hydropolitics of crisis in Chile. *Journal of Latin American Geography, 22*(3), 140–150. https://doi.org/10.1353/lag.2023.a915671

Assis Dias, M. C. D., Saito, S. M., Alvalá, R. C. D. S., Stenner, C., Pinho, G., Nobre, C. A., Fonseca, M. R. D. S., Santos, C., Amadeu, P., Silva, D., Lima, C. O., Ribeiro, J., Nascimento, F., & Corrêa, C. D. O. (2018). Estimation of exposed population to landslides and flood risk areas in Brazil, on an intra-urban scale. *International Journal of Disaster Risk Reduction, 31*, 449–459. https://doi.org/10.1016/j.ijdrr.2018.06.002

Azócar, G., Billi, M., Calvo, R., Huneeus, N., Lagos, M., Sapiains, R., & Urquiza, A. (2021). Climate change perception, vulnerability, and readiness: Inter-country variability and emerging patterns in Latin America. *Journal of Environmental Studies and Sciences, 11*(1), 23–36. https://doi.org/10.1007/s13412-020-00639-0

Bachelard, G. (1989). *A poética do espaço*. Martins Fontes.

Barrales Díaz, C. (2019). Atención psicológica en situaciones de emergencias y desastres. *Horizonte Sanitario, 18*(1), 5–6.

Berroeta, H., & Pinto de Carvalho, L. (2020). La psicología ambiental-comunitaria en el estudio de los desastres: La importancia de los vínculos socioespaciales. *Psykhe, 29*(1), 1–16. https://doi.org/10.7764/psykhe.29.1.1579

Blaikie, P., Cannon, T., Davis, I., & Wisner, B. (1994). *At risk: Natural hazards, people's vulnerability, and disasters*. Routledge.

Boström, M., & Lidskog, R. (2024). *Environmental Sociology and Social Transformation: Key Issues*. Routledge.

Bourdieu, P. (1991). *Language and symbolic power*. Polity Press.

Bradshaw, S. (2024). Extractivism and the engendering of disasters: Disaster risk creation in the era of the Anthropocene. *Disaster Prevention and Management, 33*(3), 194–205. https://doi.org/10.1108/DPM-06-2023-0146

Branco, L. L. C. (2023). Rethinking governance through Samarco's dam collapse in Brazil: A critique from the STS perspective. *Tapuya: Latin American Science, Technology and Society, 6*(1). https://doi.org/10.1080/25729861.2022.2162708

Brasil. (2004). Relatório Sala de Situação—Enchentes 2004. Sedec.

Bravo, L., & Fragkou, M. C. (2019). Escasez hídrica, género, y cultura mapuche: Un análisis desde la ecología política feminista. *Polis. Revista Latinoamericana,* (54). http://journals.openedition.org/polis/17893

Bravo, L. A. (2022). *A governança do sistema federal de alerta de risco de desastres associados a deslizamentos no Brasil* (Master's thesis, Universidade Estadual Paulista). http://hdl.handle.net/11449/234653

Brenes, A., Marchezini, V., & Wisner, B. (2020, January 30). Non-governmental organizations and natural hazard governance in Latin America and the Caribbean. In *Oxford research encyclopedia of natural hazard science.* Retrieved September 5, 2025, from https://oxfordre.com/naturalhazardscience/view/10.1093/acrefore/9780199389407.001.0001/acrefore-9780199389407-e-333

Bustos, B., Prieto, M., & Barton, J. (2015). *Ecología política en Chile: Naturaleza, propiedad, conocimiento y poder*. Universitaria.

Cadierno-Gutiérrez, J. (2018). Transmisión de conocimiento ancestral lavkenche sobre terremotos y tsunamis y su implicancia en la reducción del riesgo de desastres. *REDER, 2*(2), 16. https://doi.org/10.55467/reder.v2i2.15

Campos, L., Suazo, V., & Cárdenas, A. (2021). Experiencias de relegación urbana post-desastre: El caso de los damnificados del gran incendio de Valparaíso 2014. *Revista CUHSO, 31*(2), 126–150. https://doi.org/10.7770/cuhso-v31n2-art2248

Carmo, F. F. do, Kamino, L. H. Y., Junior, R. T., Campos, I. C. de, Silvino, G., Castro, K. J. da S. X. de, Mauro, M. L., Rodrigues, N. U. A., Miranda, M. P. de S., & Pinto, C. E. F. (2017). Fundão tailings dam failures: The environment tragedy of the largest technological disaster of Brazilian mining in global context. *Perspectives in Ecology and Conservation, 15*(3), 145–151. https://doi.org/10.1016/J.PECON.2017.06.002

Couto, G. A. (2023). *Climate related disasters in Brazil through a gender lens: Sobrevidas@ and female voices* (Doctoral dissertation, National Institute for Space Research, Doctorate Program on Earth System Science). http://mtc-m21d.sid.inpe.br/col/sid.inpe.br/mtc-m21d/2023/03.10.18.53/doc/publicacao.pdf

Eriksen, C., Kirschner, J., Simon, G. L., O'Grady, N., Uyttewaal, K., Lüthi, S., Prior, T., Zeffiri, F., Emmenegger, R., Ay, D., Chmutina, K., Raju, E., & Grove, K. (2025). From rigidity traps towards reparative disaster governance and management. *International Journal of Disaster Risk Reduction, 125*, Article 105603. https://doi.org/10.1016/j.ijdrr.2025.105603

Faas, A. J., & Barrios, R. (2015). Applied anthropology of risk, hazards, and disasters. *Human Organization, 74*(4), 287–295.

Fernandes, G. W., Goulart, F. F., Ranieri, B. D., Coelho, M. S., Dales, K., Boesche, N., Bustamante, M., Carvalho, F., Carvalho, D., Dirzo, R., Fernandes, S., Galetti, P., Jr., Millan, V. G., Mielke, C., Ramirez, J., Neves, A., Rogass, C., Pinheiro, S., Scariot, A., & Soares-Filho, B. (2016). Deep into the mud: Ecological and socio-economic impacts of the dam breach in Mariana. *Brazil. Natureza & Conservação, 14*(2), 35–45.

Fitz-Henry, E., & Klein, E. (2024). From just transitions to reparative transformations. *Political Geography, 108*, Article 103004. https://doi.org/10.1016/j.polgeo.2023.103004
Fritz, C. E. (1961). Disaster. In R. K. Merton & R. A. Nisbet (Eds.), *Contemporary social problems* (pp. 651–694). Harcourt.
Gaillard, J. C. (2022). The invention of disaster: Power and knowledge in discourses on hazards and vulnerability. Routledge.
Gantus-Oliveira, T. (2024). Vulnerabilidade de gênero e raça e o olhar interseccional sobre os desastres. *Revista Estudos Feministas, 32*(1), Article e92823. https://doi.org/10.1590/1806-9584-2024v32n192823
García, M. (2022). Cultura de la prevención en desastres: Implicancias para la resiliencia social en comunidades vulnerables. *Revista Latinoamericana de Ciencias Sociales, 28*(1), 101–119. https://doi.org/10.24215/18522101.28.e10
García-Acosta, V. G. (Ed.). (1996). *Historia y desastres en América Latina I*. Centro de Investigaciones y Estudios Superiores en Antropología Social.
García-Acosta, V. G. (Ed.). (1997). *Historia y desastres en América Latina I*. Centro de Investigaciones y Estudios Superiores en Antropología Social.
García-Acosta, V. G. (Ed.). (2008). *Historia y desastres en América Latina I*. Centro de Investigaciones y Estudios Superiores en Antropología Social.
García-Acosta, V. G. (2020). Introduction: Anthropologists studying disaster in Latin America: Why, when, how? In V. García-Acosta (Ed.), The anthropology of disasters in Latin America: State of the art (pp. 1–21). Routledge.
Geraldi, D. (2009). Pessoas com deficiência visual: Do estigma às limitações da vida cotidiana em circunstâncias de riscos e de desastres relacionados às chuvas. In N. Valencio, M. Siena, V. Marchezini, & J. C. Gonçalves (Eds.), Sociologia dos desastres: construção, interfaces e perspectivas no Brasil (pp. 107–118). Rima.
Girot, P. (2020). *Desastres y vulnerabilidad en la era del cambio climático: Retos y respuestas para América Latina*. Editorial Pontificia Universidad Javeriana.
González, J. S., & Pérez, F. M. (2021). El papel de las organizaciones no gubernamentales en la gestión de desastres naturales. *Revista Mexicana de Sociología, 83*(4), 691–713. https://doi.org/10.22201/iis.01871260.2021.83.4.42637
González-Muzzio, C., & Sandoval, V. (2018). Capitalismo de desastre en procesos de reconstrucción en Chile. In J. Inzulza, A. Maragaño, C. Boano, & I. Diaz (Eds.), *Reconstrucción de ciudades intermedias en el siglo XXI: Procesos de gentrificación post desastres naturales* (pp. 41–52). Editorial Universidad de Talca.
Hamdanieh, L., Stephens, C., Olyaeemanesh, A., & Ostadtaghizadeh, A. (2024). Social justice: The unseen key pillar in disaster risk management. *International Journal of Disaster Risk Reduction, 101*, Article 104229. https://doi.org/10.1016/j.ijdrr.2023.104229
Hernández, R., Fernández, C., & Baptista, P. (2019). *Metodología de la investigación*. McGraw-Hill.
Horton, T., & MacLean, K. (2020). Climate change adaptation in the Global South: An analysis of strategies in South America. *Environmental Science & Policy, 114*, 19–29. https://doi.org/10.1016/j.envsci.2020.07.001

Hummell, B. M. L., Cutter, S. L., & Emrich, C. T. (2016). Social vulnerability to natural hazards in Brazil. *International Journal of Disaster Risk Science, 7*(2), 111–122. https://doi.org/10.1007/s13753-016-0090-9

Ibarra, M. I., & Riquelme, W. (2019). Sentipensar mapuche con las aguas del Huenehue: Hacia una ecología política y una antropología por demanda. *Polis. Revista Latinoamericana,* (54), 56–74. https://doi.org/10.32735/S0718-6568/2019-N54-1402

Iglesias, S., & Rodríguez, D. (2022). Resilience and sustainable reconstruction: A case study of the 2010 Chile earthquake. *International Journal of Disaster Risk Reduction, 61,* Article 102295. https://doi.org/10.1016/j.ijdrr.2021.102295

Imperiale, A. J., & Vanclay, F. (2024). From project-based to community-based social impact assessment: New social impact assessment pathways to build community resilience and enhance disaster risk reduction and climate action. *Current Sociology*. https://doi.org/10.1177/00113921231203168

Inostroza-Matus, C. C., Molina-Camacho, F., & Romero-Toledo, H. (2019). Desde la amenaza natural al desastre: Una construcción histórica del terremoto y tsunami de 1960 en Saavedra. *Íconos: Revista de Ciencias Sociales, 66*(1), 31–57. https://doi.org/10.17141/iconos.66.2020.4016

Instituto Brasileiro do Meio Ambiente e dos Recursos Naturais Renováveis (IBAMA). (2015). *Impactos ambientais decorrentes do desastre envolvendo o rompimento da barragem de Fundão, em Mariana, Minas Gerais.* http://www.ibama.gov.br/phocadownload/noticias%20ambientais/laudotecnico%20preliminar%20Ibama.pdf

Inzunza, S. I., Castañeda, J., & Carraro, V.. (2023). Resiliencia comunitaria en contexto de desastre por COVID-19: Resurgimiento y territorialización de las ollas comunes, el caso de Puente Alto. *Revista de Geografía Norte Grande, 84,* 245–270. https://doi.org/10.4067/s0718-34022023000100245

Klein, N. (2007). *The shock doctrine: The rise of disaster capitalism.* Metropolitan Books.

Lampis, A., Brink, E., & dos Santos, A. H. (2022). Reparation ecology and climate risk in Latin America: Experiences from four countries. *Frontiers in Climate, 4,* Article 897424. https://doi.org/10.3389/fclim.2022.897424

Lavell, A. (1993). Ciencias sociales y desastres naturales en América Latina: Un encuentro inconcluso. In A. Maskrey (Ed.), *Los desastres no son naturales* (pp. 111–125). La Red.

Lavell, A. (2005). Los conceptos, estudios y práctica en torno al tema de los riesgos y desastres en América Latina: Evolución y cambio, 1980–2004: El rol de la Red, sus miembros y sus instituciones de apoyo. In *La gobernabilidad en América Latina: Balance reciente y tendencias a futuro* (pp. 1–66). Facultad Latinoamericana de Ciencias Sociales. https://goo.gl/5TydaR

Lindell, M. (2011). Disaster studies. In *Sociopedia.isa*. Sage. https://www.academia.edu/30969496/Disaster_studies

Losekann, C. (2017). "It was no accident!" The place of emotions in the mobilization of people affected by the collapse of Samarco's tailings dam in Brazil. *Vibrant:*

Virtual Brazilian Anthropology, 14(2), e142102. https://doi.org/10.1590/1809-43412017v14n2p102

Losekann, C., & Milanez, B. (2023). Mining disaster in the Doce River: Dilemma between governance and participation. *Current Sociology*. https://doi.org/10.1177/00113921211059224

Madariaga, A. (2020). The three pillars of neoliberalism: Chile's economic policy trajectory in comparative perspective. *Contemporary Politics, 26*(3), 308–329. https://doi.org/10.1080/13569775.2020.1735021

Marchezini, V. (2010). *Desafios de gestão de abrigos temporários: Uma análise sociológica de inseguranças e riscos no cotidiano de famílias abrigadas (Master's thesis*. Universidade Federal de São Carlos.

Marchezini, V. (2013). *Janeiro de 2010, São Luiz do Paraitinga/SP: Lógicas de poder, discursos e práticas em torno de um desastre (Doctoral thesis*. Universidade Federal de São Carlos.

Marchezini, V. (2014). La producción silenciada de los "desastres naturales" en catástrofes sociales. *Revista Mexicana de Sociología, 76*(2), 253–285. http://www.scielo.org.mx/scielo.php?script=sci_arttext&pid=S0188-25032014000200004&lng=es&tlng=es

Marchezini, V. (2015). The biopolitics of disaster: Power, discourses, and practices. *Human Organization, 74*(4), 362–372. https://www.jstor.org/stable/26536814

Marchezini, V. (2018). As ciências sociais nos desastres: Um campo de pesquisa em construção. *Revista Brasileira De Informação Bibliográfica Em Ciências Sociais, 83*, 43–72.

Marchezini, V. (2019). The power of localism during the long-term disaster recovery process. *Disaster Prevention and Management, 28*(1), 143–152. https://doi.org/10.1108/DPM-05-2018-0150

Marchezini, V. (2020). "What is a sociologist doing here?" An unconventional people-centered approach to improve warning implementation in the Sendai framework for disaster risk reduction. *International Journal of Disaster Risk Science, 11*, 218–229. https://doi.org/10.1007/s13753-020-00262-1

Marchezini, V. (2024). Why are you in disaster studies? Liberating future scholars from oppressive disaster science. *Disaster Prevention and Management, 33*(3), 223–236. https://doi.org/10.1108/DPM-06-2023-0150

Marchezini, V., & Wisner, B. (2017). Challenges for vulnerability reduction in Brazil: Insights from the PAR framework. In V. Marchezini, B. Wisner, L. Londe, & S. Saito (Eds.), *Reduction of vulnerability to disasters: From knowledge to action* (pp. 57–96). Rima Editora.

Marchezini, V., Saito, S. M., Londe, L. R., & Damacena, F. D. L. (2025). Implementation challenges of disaster risk management policies: The organizational capacities of municipal civil defense units. *International Journal of Disaster Risk Reduction, 119*, Article 105291. https://doi.org/10.1016/j.ijdrr.2025.105291

Marchezini, V., Trajber, R., Olivato, D., et al. (2017). Participatory early warning systems: Youth, citizen science, and intergenerational dialogues on disaster risk reduction in Brazil. *International Journal of Disaster Risk Science, 8*, 390–401. https://doi.org/10.1007/s13753-017-0150-9

Marchezini, V., Maciel Ferreira, M. R., & Severo, F. (2026). Resisting necropolitics: Mutual aid and grassroots responses to the COVID-19 syndemic in Brazil. In R. E. Barrios & V. García-Acosta (Eds.), *A Revelatory Pandemic: Crisis, Agency, and COVID in Latin America* (pp. 118–141). Berghahn Books. https://doi.org/10.2307/jj.29556569.11

Marengo, J. A., Jimenez, J. C., Espinoza, J., Cunha, A. P., & Aragão, L. E. (2022). Increased climate pressure on the agricultural frontier in the Eastern Amazonia-Cerrado transition zone. *Scientific Reports, 12*(1), 1–10. https://doi.org/10.1038/s41598-021-04241-4

Martínez, C., Cienfuegos, R., Inzunza, S., Urrutia, A., & Guerrero, N. (2020). Worst-case tsunami scenario in Cartagena Bay, central Chile: Challenges for coastal risk management. *Ocean & Coastal Management, 185*, Article 105060. https://doi.org/10.1016/j.ocecoaman.2019.105060

Maskrey, A. (1993). *Disasters are not natural (Los desastres no son naturales).* La Red. https://www.desenredando.org/public/libros/1993/ldnsn/LosDesastresNoSonNaturales-1.0.0.pdf

Mattedi, M. A. (1999). *As enchentes como tragédias anunciadas: Impactos da problemática ambiental nas situações de emergência em Santa Catarina (Doctoral thesis.* Universidade Estadual de Campinas.

Matthewman, S., & Lambert (Tūhoe, Ngāti Ruapani), S. J. (2024). Standing upright here: Critical disaster studies viewed from the Antipodes. *Kōtuitui: New Zealand Journal of Social Sciences Online, 20*(2), 143–158. https://doi.org/10.1080/1177083X.2024.2345767

Mbembe, A. (2019). *Necropolitics.* Duke University Press.

Mena, C. (2018). ¿Cambio con transformación? Políticas públicas de desastre en Chile post Terremoto 2010, un caso de estudio. *REDER, 2*(1), 4–20.

Micheletti, S., & Letelier Troncoso, F. (2016). Damnificados de la reconstrucción post-terremoto: Efectos del modelo en el hábitat rural del Maule. *Revista INVI, 31*(86), 17–58. https://doi.org/10.4067/s0718-83582016000100002

Mondaca, C., & Díaz, A. (2022). Vaticinios de eventos climáticos en comunidades aymaras en los Andes del norte de Chile. *Revista De Geografía Norte Grande, 83*, 51–66. https://doi.org/10.4067/S0718-34022022000300051

Moreno, J., & Shaw, D. (2018). Women's empowerment following disaster: A longitudinal study of social change. *Natural Hazards, 92*(1), 205–224. https://doi.org/10.1007/s11069-018-3204-4

Moreno, J., & Shaw, D. (2019). Community resilience to power outages after disaster: A case study of the 2010 Chile earthquake and tsunami. *International Journal of Disaster Risk Reduction, 34*, 448–458. https://doi.org/10.1016/j.ijdrr.2018.12.016

Muniz, R. C., Ferradas, F. M., Gomez, G. M., & Pegler, L. J. (2021). Covid-19 in Brazil in an era of necropolitics: Resistance in the face of disaster. Disasters, 45, S97–S118. https://doi.org/10.1111/disa.12528

Oda, P. S. S., Marchezini, V., Lotta, G. S., Mota Ferreira, A., Cotting, A. L. M., Cota Dias, K. G., & Pacheco Calderon, O. L. (2025). State, institutional and organizational capacities in disaster risk management. *International Journal of Disaster Risk Reduction, 129*, Article 105777. https://doi.org/10.1016/j.ijdrr.2025.105777

Oliver-Smith, A. (1994). Reconstrucción después del desastre: Una visión general de secuelas y problemas. In A. Lavell (Ed.), *Al norte del Río Grande* (pp. 25–40). La Red.

Oliver-Smith, A. (2012). The brotherhood of pain: Theoretical and applied perspectives on post-disaster solidarity. In A. Oliver-Smith & S. M. Hoffman (Eds.), *The angry earth: Disaster in anthropological perspective* (pp. 204–225). Routledge.

Osorio-Parraguez, P., & Espinoza, A. (2016). Salud mental en desastres naturales: Estrategias interventivas con adultos mayores en sectores rurales de Chile. *Global Health Promotion, 23*(2), 84–91. https://doi.org/10.1177/1757975914566890

Panez, A. (2022). *El río recuperando su cauce: Despojos y resistencias en los conflictos por agua-tierra-territorio bajo el neoliberalismo en Chile.* EDUEPB.

Peek, L., & Guikema, S. (2021). Interdisciplinary theory, methods, and approaches for hazards and disaster research: An introduction to the special issue. *Risk Analysis, 41*(7), 1047–1058. https://doi.org/10.1111/risa.13777

Peek, L., Tobin, J., Adams, R. M., Wu, H., & Mathews, M. C. (2020). A framework for convergence research in the hazards and disaster field: The Natural Hazards Engineering Research Infrastructure CONVERGE facility. *Frontiers in Built Environment, 6*, Article 551252. https://doi.org/10.3389/fbuil.2020.00110

Peek, L., Wachtendorf, T., & Meyer, M. A. (2021). Sociology of disasters. In B. Schaefer Caniglia, A. Jorgenson, S. A. Malin, L. Peek, D. N. Pellow, & X. Huang (Eds.), *Handbook of environmental sociology* (Handbooks of Sociology and Social Research). Springer. https://doi.org/10.1007/978-3-030-77712-8_11

Peek, L., Mark, B., Austin, J., & Champeau, H. (2024). 2023 Social Science Extreme Events Research (SSEER) Network. In *Social Science Extreme Events Research (SSEER) Network data, survey instrument, and annual census.* DesignSafe-CI. https://doi.org/10.17603/ds2-584f-vz32

Perry, R. W. (2005). Disasters, definitions and theory construction. In R. W. Perry & E. L. Quarantelli (Eds.), *What is a disaster? New answers to old questions* (pp. 311–324). Xlibris.

Perry, R. W., & Quarantelli, E. L. (Eds.). (2005). *What is a disaster? New answers to old questions.* Xlibris.

Phillips, B. (2014). *Qualitative disaster research.* Oxford University Press.

Pismel, G. O., Marchezini, V., Selaya, G., de Paula, Y. A., Mendoza, E., & Anderson, L. O. (2023). Wildfire governance in a tri-national frontier of southwestern Amazonia: Capacities and vulnerabilities. *International Journal of Disaster Risk Reduction, 86*, Article 103529. https://doi.org/10.1016/j.ijdrr.2023.103529

Posner, P. W. (2024). Social welfare policy in post-transition Chile: Social democratic or neoliberal? *Critical Sociology, 50*(7–8), 1207–1222. https://doi.org/10.1177/08969205231209424

Prince, S. H. (1920). *Catastrophe and social change based upon a sociological study of the Halifax disaster* (Doctoral dissertation, Columbia University).

Quarantelli, E. L. (1998). *What is a disaster?* Routledge.

Quarantelli, E. L. (2006). *Catastrophes are different from disasters: Some implications for crisis planning and managing drawn from Katrina.* Social Science Research Council. https://items.ssrc.org/understanding-katrina/catastrophes-are-different-from-disasters-some-implications-for-crisis-planning-and-managing-drawn-from-katrina/

Ramos-Chocobar, S., & Tironi, M. (2023). Un Sol Interior: Vulcanología Lick-anantay en el Salar de Atacama. *REDER, 7*(2), 73. https://doi.org/10.55467/reder.v7i2.125

República de Chile. (2021). *Establece el Sistema Nacional de Prevención y Respuesta ante Desastres, sustituye la Oficina Nacional de Emergencia por el Servicio Nacional de Prevención y Respuesta ante Desastres, y adecúa normas que indica. Diario Oficial de la República de Chile.*

Ribeiro, M. W. T. (2020). O encontro das antropologias do meio ambiente e dos desastres no Brasil. *BIB: Revista Brasileira de Informação Bibliográfica em Ciências Sociais,* (93), 1–25. https://bibanpocs.emnuvens.com.br/revista/article/view/517

Rivas, R.. (2022). La posteridad de la crisis: Notas sobre las contradicciones socioecológicas de la industria del salmón en Chile. *Cuadernos de Teoría Social, 7*(14), 74–86. https://doi.org/10.32995/0719-64232021v7n14-123

Rivas, R., Acuña, V., & Fuentealba, R. (2024). Hacia una reflexión crítica de los desastres socionaturales en Chile. *Polis. Revista Latinoamericana, 23*(67), 1–8. https://doi.org/10.32735/S0718-6568/2024-N67-3640

Rojas, C. (2020). O rompimento da barragem de rejeitos de Fundão e a ascensão do capitalismo de desastre no Brasil [Doctorate thesis, Universidade Federal de Minas Gerais]. Institutional Repository of Universidade Federal de Minas Gerais. https://repositorio.ufmg.br/bitstreams/f91c2657-599a-41f1-ad68-e58065c65dd2/download

Rodrigues, A. C., Gunther, W. M. R., Vasconcellos, M. P., Di Giulio, G. M., & Boscov, M. E. G. (2015). Delineamento da produção científica sobre desastres no Brasil no início deste século. *Desenvolvimento e Meio Ambiente, 34*, 61–73.

Romero-Aravena, H., & Mendes, F. H. (2021). La construcción social de climas urbanos y su relación con la pandemia de Covid-19 en Santiago de Chile. *Cuadernos de Geografía: Revista Colombiana de Geografía, 30*(2), 376–395. https://doi.org/10.15446/rcdg.v30n2.88701

Romero Toledo, H., & Romero Aravena, H. (2015). Ecología política de los desastres: Vulnerabilidad, exclusión socio-territorial y erupciones volcánicas en la Patagonia chilena. *Magallania, 43*(3), 7–26.

Saavedra, J. (2023). Biopolítica del desastre e intervención social en la ciudad. *Propuestas Críticas En Trabajo Social, 3*(6), 77–99. https://doi.org/10.5354/2735-6620.2023.70054

Saavedra, J., Azócar, G., & Fernández-Vicente, P. (2021). Neoliberalism of disaster and long-term recovery: The case of the 2010 earthquake in Talcahuano, Chile.

International Journal of Disaster Risk Reduction, 61, Article 102356. https://doi.org/10.1016/j.ijdrr.2021.102356

Sandoval, V., González-Muzzio, C., & Castro, C. P. (2024). El poder de una re-imaginación radical de los desastres: Nexo poscolonial y capitalista en Chile y América Latina (e ideas para un futuro justo y equitativo). *REDER, 8*(2), 241. https://doi.org/10.55467/reder.v8i2.170

Sandoval, V., González-Muzzio, C., Villalobos, C., Sarmiento, J. P., & Hoberman, G. (2020). Assessing disaster capitalism in post-disaster processes in Chile: Neoliberal reforms and the role of the corporate class. *Disaster Prevention and Management, 29*(6), 831–847. https://doi.org/10.1108/DPM-01-2020-0005

Sandoval, V., & Voss, M. (2016). Disaster governance and vulnerability: The case of Chile. *Politics and Governance, 4*(4), 107–116. https://doi.org/10.17645/pag.v4i4.743

Sandoval, V., & Voss, M. (2018). Multi-scalar disaster governance and vulnerability in Chile. In G. Forino, S. Bonati, & L. Calandra (Eds.), *Governance of risk, hazards and disasters* (pp. 27–42). Routledge.

Schuller, M. (2008). Deconstructing the disaster after the disaster: Conceptualizing disaster capitalism. In N. Gunewardena & M. Schuller (Eds.), *Capitalizing on catastrophe: Neoliberal strategies in disaster reconstruction* (pp. 17–27). AltaMira Press.

Sepúlveda, E. (2024). Socio-educational practices of community environmental educators from territories affected by socio-environmental conflicts and socio-natural disasters in Chile: Roles, ethics, and methodologies. *The Journal of Environmental Education*, 1–14. https://doi.org/10.1080/00958964.2024.2364179

Siena, M. (2009). *A dimensão de gênero na análise sociológica dos desastres: Conflitos entre desabrigadas e gestoras de abrigos temporários relacionados às chuvas* (Master's thesis, Universidade Federal de São Carlos).

Soares-Guimarães, A., Cardoso, F., & Wanderley, F. (2025). Naturaleza y política en la América Latina contemporánea. *Debates en Sociología, 60*, 8–16. https://doi.org/10.18800/debatesensociologia.202501.000

Sorokin, P. (1942). *Man and society in calamity: The effects of war, revolution, famine, pestilence upon human mind, behavior, social organization and cultural life*. E. P. Dutton.

Taddei, R. R. (2014). Sobre a invisibilidade dos desastres na antropologia brasileira. *Waterlat-Gobacit Network Working Papers, 1*(1), 30–42.

Thomalla, F., Boyland, M., Johnson, K., Ensor, J., Tuhkanen, H., Gerger Swartling, Å., Han, G., Forrester, J., & Wahl, D. (2018). *Transforming development and disaster risk. Sustainability, 10*(5), 1458. https://doi.org/10.3390/su10051458

Tironi, M. (2022). Hesitant: Three theses on ecological reparation (otherwise). In D. Papadopoulos, M. Puig de la Bellacasa, & M. Tacchetti (Eds.), *Ecological reparation: Repair, remediation, and resurgence in social and environmental conflict* (pp. xx–xx). Bristol University Press.

Tironi, M., & Undurraga, B. (2023). Provocaciones y tensiones del Antropoceno. *Estudios Públicos, (171)*, 137–179. https://doi.org/10.38178/07183089/1936210623

Turner, B. (2023). Towards a sociology of catastrophe: The case of Covid-19. In J. M. Ryan (Ed.), *Covid-19: Surviving a pandemic* (pp. 26–37). Routledge. https://doi.org/10.4324/9781003302698

Ubilla-Bravo, G., & Rodríguez-Seguel, V. (2022). *Asentamientos humanos en Chile: Revisión general en torno al cambio climático y ambiental para ciudades grandes e intermedias. Estado de la materia en 2022* (Research report, p. 50). https://doi.org/10.5281/zenodo.7051295

Valencio, N. (2012). *Para além do "dia do desastre" – o caso brasileiro.* Appris.

Valencio, N. F. L. S., Siena, M., & Marchezini, V. (2011). *Abandonados nos desastres: Uma análise sociológica de dimensões objetivas e simbólicas de afetação de grupos sociais desabrigados e desalojados* (1st ed.). Conselho Federal de Psicologia.

Vasconcelos Junior, F. D. C., Zachariah, M., Silva, T. L. D. V., Santos, E. P. D., Coelho, C., Alves, L. M., Martins, E. S. P. R., Köberle, A. C., Singh, R., Vahlberg, M., Marchezini, V., Heinrich, D., Thalheimer, L., Raju, E., Koren, G., Philip, S. Y., Kew, S. F., Bonnet, R., Li, S., & Otto, F. (2024). An attribution study of very intense rainfall events in Eastern Northeast Brazil. *Weather and Climate Extremes, 45*, Article 100699. https://doi.org/10.1016/j.wace.2024.100699

Vergara, P., & Ejsmentewicz, D. (2024). Implementación de nueva institucionalidad para la reducción del riesgo de desastres en Chile: Reflexiones sobre los avances y desafíos para la gestión del riesgo de desastres. *Polis. Revista Latinoamericana, 23*(67), 9–57. https://doi.org/10.32735/S0718-6568/2024-N67-3626

Vergara, P., Fuster-Farfán, X., & Miranda, F. (2022). Desigualdades en contextos de desastres socionaturales: Reflexiones desde el habitar interseccional de mujeres lideresas. *Revista INVI, 37*(104), 71–99. https://doi.org/10.5354/0718-8358.2022.65947

Vieira, R. M. D. S. P., Sestini, M. F., Tomasella, J., Marchezini, V., Pereira, G. R., Barbosa, A. A., Santos, F. C., Rodriguez, D. A., do Nascimento, F. R., Santana, M. O., Barreto Campello, F. C., & Ometto, J. P. H. B. (2020). Characterizing spatio-temporal patterns of social vulnerability to droughts, degradation and desertification in the Brazilian northeast. *Environmental and Sustainability Indicators, 5*, 100016. https://doi.org/10.1016/j.indic.2019.100016

Wisner, B., Alcántara-Ayala, I., Gaillard, J.C., Kelman, I., Marchezini, V. (2025). Understanding and addressing disaster risk: Who speaks? Who suffers? (1st ed.). Routledge. https://doi.org/10.4324/9781003292814

Wisner, B., Gaillard, J.-C., & Kelman, I. (2004). *At risk: Natural hazards, people's vulnerability, and disasters*. Routledge.

Wisner, B., Gaillard, J.C., & Kelman, I. (2012). Framing disaster: Theories and stories seeking to understand hazards, vulnerability, and risk. In B. Wisner, J.C. Gaillard, & I. Kelman (Eds.), *The Routledge handbook of hazards and disaster risk reduction* (pp. 18–34). Routledge.

Zhouri, A., Oliveira, R., Zucarelli, M., & Vasconcelos, M. (2017). The Rio Doce mining disaster in Brazil: Between policies of reparation and the politics of affectation. *Vibrant: Virtual Brazilian Anthropology, 14*(2), e142081. https://doi.org/10.1590/1809-43412017v14n2p081

6

Metabolism

Debra J. Davidson and Steven Yearley

Metabolism Narratives in Environmental Studies

In use by scientists for centuries, the concept of metabolism originally referred to physical and chemical processes operating at the cellular level (Baccini & Brunner, 2023), but the concept has been usefully applied to organisms, ecosystems, human communities, and even economic sectors. At its core, metabolism refers to the efficiency with which an organism takes up and utilizes the energy and nutrients it requires to maintain function, including "organisms" as large and complex as societies.

Metabolism has been, and can continue to serve as, a central concept for understanding the co-relations between social- and ecosystems. Use of the concept of metabolism in studies of socio-ecological relations has continued to grow during the first decades of the twenty-first century (Haberl et al., 2019), and thus we devote a chapter to this important concept once again.

As described by Haberl and colleagues (2019: 173), "sociometabolic research (SMR) is a systems approach to studying society–nature interactions at different spatiotemporal scales [which] encompasses biophysical flows exchanged between societies and their natural environment as well as the

D. J. Davidson (✉)
University of Alberta, Edmonton, AB, Canada
e-mail: debra.davidson@ualberta.ca

S. Yearley
University of Edinburgh, Edinburgh, UK
e-mail: steve.yearley@ed.ac.uk

D. J. Davidson et al. (eds.), *Environment and Society*,
https://doi.org/10.1007/978-3-032-27059-7_6

flows within and between social systems." Metabolism thus draws attention to relationships between a given organism and its environment: including the uptake of energy, nutrients and resources from ecosystems, their efficient distribution through that organism to where they are needed—for example, the brain or the circulatory system—and the release by the organism of wastes back into the environment. Insufficiency of required inputs to sustain the organism, the breakdown within the organism of the ability to utilize those inputs or excrete wastes, or the overexploitation of the environment such that availability of inputs or absorptive capacity of wastes declines, will compromise the survival of that organism.

The concept's relevance to socio-ecological systems is not hard to see. Metabolism draws our attention to three primary necessities of a socio-ecological system, whether that system is a family farm or the global economy: access to adequate inputs to support the system, the presence of mechanisms within that system to process and distribute those inputs to where they are needed, and the need to avoid overexploitation of the environment's capacity to absorb wastes. The larger and more complex the system, the larger the volume of inputs and more sophisticated the processing and distribution mechanisms will tend to be. In other words, an economic system that does not receive sufficient energy, material inputs, or financial investments will be unstable. But metabolism also tells us that an economic system that does receive sufficient inputs, but does not ensure mechanisms for the *distribution* of those inputs to support the system as a whole—such as when wealth becomes concentrated among a small elite group of oligarchs thereby supporting a highly inequitable economic structure, or when there is inadequate infrastructure to convert energy sources into reliable electricity—those systems will likewise be unstable. Metabolism is also understood to be dynamic; just as one's individual biophysical metabolism changes as we age, so too do the metabolic requirements of social systems and individuals living within those systems.

Baccini and Brunner (2023: 28), for example, highlight the fact that the metabolic requirements of an individual living in a modern society are magnitudes higher than, say, a hunter-gatherer. In other words, the "anthropogenic metabolism of a person is at least 10 times larger than the physiologic metabolism and comprises all goods needed to reside, to work, to communicate, to eat, to breathe, to clean, and so forth." They elaborate this argument with an in-depth depiction of the metabolism of a contemporary Global North household, encompassing everything from water, energy and consumables to the many forms of resulting waste flows. As such, Baccini and Brunner (2023: 29, emphasis added) continue, incidentally offering a compelling rebuttal to arguments that the number of greedy humans is the primary source of our ecological ills: "the physiologic metabolism of a human *individual* is not the key process of the anthroposphere. It is the anthropogenic metabolism of a human *society*."

In this chapter we explore this anthropogenic metabolism, beginning with a brief historical tour of metabolism's conceptual development and applications. This is followed by an elaboration of this discussion of metabolism into two arenas with specific relevance to Just Transformation, beginning with a discussion of the recent rise to prominence of discourses seeking to inform the transformation of capitalist economies to support a more reciprocal socio-ecological metabolism, with attention to one such discursive thread in particular: circular economies. Second, we turn our attention to collective action: the means by which metabolic rifts can create conditions for social mobilization, and furthermore, how a metabolic lens can help us identify some essential sources of sustenance to enable the emergence, and persistence, of social movements to support Just Transformation.

History of a Concept

While the concept's use dates back over four centuries, its use has been revitalized and expanded upon during the past three decades due to its unique utility for understanding contemporary crises in our socio-ecological relations, and it is this recent history to which we briefly turn here. Initial efforts to articulate a socio-ecological metabolism include natural scientists such as Howard T. and Eugene Odum (1969, 1995, 2007; Odum & Odum, 2001). The Odums were among the first scholars to apply the metabolism concept to biological levels above the organism, including societies, with their development of "emergy," intended to serve as a common metric to capture the multitude of inputs required to provide societies with access to energy (and thus the generation of wealth). This work was a precursor to Energy Return on Investment (EROI) analyses, which calculate the amount of energy lost during its transformation into forms of energy that people can use. In the social sciences, Leslie White and Frederick Cottrell (Cottrell, 1955; White, 1943) both shared the Odums' focus on the energy requirements of social systems and how energy access—and in particular, the technologies employed to do so—determines the rate and form of economic development, and ultimately culture.

The concept has been used, both implicitly and explicitly, to describe cities in particular, beginning with researchers working with Robert E. Park at the University of Chicago in the 1920s. Park and his colleagues sought to describe cities as ecological systems that thrive as a function of material stocks and flows (Park et al., 1925). It was not until four decades later, however, that the term "urban metabolism" was used, introduced by an engineer, Abel Wolman. Wolman defined the city's metabolic needs as "all the materials and commodities needed to sustain the city's inhabitants at home, at work and at play" (Wolman, 1965: 179). Although written before widespread awareness of the global scale of many environmental problems, a key contribution of this scholarship was attention to waste flows: the pollution that results from urbanization.

It took a dramatic increase in computing power to fully articulate the contours of urban metabolism, which a group of researchers at the Autonomous University of Spain in Barcelona have utilized to support their Multi-scale Integrated Analysis of Societal and Ecosystem Metabolism (MuSIASEM) (Giampetro et al., 2009; Ramos-Martín et al., 2007, 2009). MuSIASEM is designed to support forecasting of the energy metabolism of alternative development pathways, including attention to the implications of resource scarcity and historic trends in energy intensity. Researchers at the Institute of Social Ecology in Vienna have also significantly enriched the field, both theoretically and empirically. A collection of research undertaken at the Institute (Fischer-Kowalski & Weisz, 2016) depicts interactions across social and ecological systems as co-evolutionary: arguing that the ability to provide work, energy and material flows for social systems requires continuous social reorganization and new interventions in ecosystems. Importantly, these researchers pay heightened attention to the role of culture and communication in these interrelationships, important drivers of change that were overlooked in most earlier metabolism analyses.

The works discussed so far have been highly productive in extending the application of the concept of metabolism to society-environment interactions, in a somewhat descriptive fashion. Another set of works, however, seeks to deploy the concept of metabolism in critical theoretical accounts of global capitalism, including, for example, Karl Polanyi, Nicholas Georgescu-Roegen, Joan Martinez-Alier, and Stephen Bunker. Polanyi (1944) was a rare figure for his era, positing that the necessary pursuit of profit by capitalists destroys the socio-ecological resources that capitalism requires in the process of creating market value, an internal mechanism that James O'Connor (1988) would later refer to as the Second Contradiction of Capital. Georgescu-Roegen (1971) highlighted the energy dependence of capitalist economic growth and the irreversible degradation of energy and material resources (entropy) in the process (See also Davidson, 2019). Martinez-Alier (2002) expanded on Georgescu-Roegen's work, attending to an even greater degree to the ecological disruptions introduced by the outputs of economic processes, and the distribution of those disruptions. Martinez-Alier also, importantly, focuses his analysis on the continuous geographic expansion of extraction activities in order to secure the energy and material requirements for capitalist growth, as exploited reserves become exhausted, in effect representing attempts to combat the inevitable decline in system metabolism posed by growing scarcity of inputs—the energy and "nutrients" required to sustain that system. The resulting separation of places of extraction from places of wealth accumulation amounts to an unequal ecological exchange, with wealth accumulating in core countries, while the environmental consequences of that wealth accumulation are "exported" to low-income countries across the periphery (Jorgenson & Clark, 2012; Rice, 2007). The ensuing ecological distribution conflicts, according to Martinez-Alier, can lead to social mobilizations in those peripheral regions, or to use his term, the "environmentalism of the poor."

Stephen Bunker's work is also well-situated along the lines of inquiry opened by Georgescu-Roegen. Bunker draws on metabolic studies provided by ecological economists, but pays particular attention to the unequal ecological exchange described by Martinez-Alier. In *Underdeveloping the Amazon* (1985), Bunker argued that the energy embodied in the raw materials extracted from peripheral regions, and the ecological waste left in its wake, served to enhance the wealth accumulation of those countries in which the materials were eventually processed and consumed, while the very bases upon which future development might ensue in the periphery are undermined. Bunker's analysis highlights the fact that the unequal exchange of labor is accompanied by the unequal exchange of matter and energy (Chase-Dunn, 1998).

Over the past 25 years, these critical metabolic analyses of capitalist economic systems have been revitalized by a handful of scholars who have revisited the writings of Karl Marx, in an effort to further explicate Marx's engagement with socio-environmental relations, and more precisely his preoccupation with how society reproduces itself, necessarily involving active engagement with nature in ways that have become non-reciprocal. For Marx (1976: 283), "Labour is, first of all, a process between man [sic] and nature, a process by which man, through his own actions, mediates, regulates and controls the metabolism between himself and nature." In other words, to be human is to engage in metabolic relation within a world governed by the laws of nature. As such, changes in the character of that metabolic relation are consequential in the same way that replacing a nutrient-rich diet with a nutrient-poor one has an effect on our biophysical condition.

What came to the fore was Marx's employment of the concept of metabolic rift. As depicted by John Bellamy Foster and his colleagues, Marx was heavily influenced by emerging insights from the soil sciences regarding the decline in soil fertility experienced over decades of modern agriculture. Marx attributed this fertility decline in large part to the "urbanization of the countryside" (Marx, 1973: 479, cited in Foster, 1999): the mass migration of people from rural, agrarian regions to cities—a movement that was elemental to capitalism. In the words of Marx again (1976: 638) this separation of town and country: "prevents the return to the soil of its constituent elements consumed by man in the form of food and clothing; hence it hinders the operation of the eternal natural condition for the lasting fertility of the soil" while simultaneously creating serious pollution problems in those city centers. The attempt to rectify declining soil fertility through the application of fertilizers, either extracted from elsewhere, such as guano from Latin America (Clark & Foster, 2009), or synthetically produced via the Haber–Bosch Process, offered temporary recovery of production on the farm, but served to exacerbate disruptions to socio-ecological metabolism in other ways. As is well-known today, fertilizer production and overuse is responsible not only for declining soil health but also for an enormous proportion of global energy consumption and by extension, greenhouse gases (Capdevila-Cortada, 2019).

Foster and colleagues take up the concept of metabolic rift as a general statement on capitalism, describing the inherent capitalist imperative to maximize the production of surplus value while minimizing production times; in turn, this demands ever-increasing volumes of materials and energy being brought into production faster than the ecological metabolic processes that allow for the replenishment of such resources (Burkett & Foster, 2006; Clark & York, 2005; Foster, 2000). As several commentaries have noted, the environmental impacts of development in state-socialist countries, industrialized East European countries and China during the Cold War were equally substantial, suggesting they are subject to the same imperative, perhaps by virtue of their engagement in the global economy (see Yearley, 1992). This trend has continued within authoritarian capitalist states such as Russia and post-Deng Xiaoping China.

As described by Kohei Saito (2023), Foster's explication of Marx's metabolic rift encompasses three dimensions. The first of these is a material rift: the fundamental disruption of cyclical material flows in ways that amount to "robbery," for example of the soil. Second, a spatial rift ensues, involving the separation of town and country, but also the imperialist absorption of non-capitalist peripheral zones into capitalism in the "race for resources" (Bunker & Ciccantell, 2005). The third dimension depicts a temporal rift, between natural time and capital time, reflecting the continuous shortening of time cycles involved in the exploitation of natural resources and production of waste; time spans that become increasingly briefer than the timespans over which resources are replenished and wastes absorbed by planetary systems. For Saito, the Anthropocene, understood as an emerging geological epoch in which humans have become the dominant driving force of nature, is a moment most aptly demarcated by the complete exhaustion of capitalism's ability to "externalize" its overexploitation of nature; the moment when nature bites back.

These various applications of the concept of metabolism to socio-ecological relations have had their critics, most poignantly the argument that the metabolic rift concept, ironically, reproduces the Cartesian dualism depicting separate natural and social spheres, and the tendency toward determinism and lack of attention to human agency (e.g. Moore, 2017). For Moore (2011: 1), "capitalism does not act upon nature so much as develop through nature–society relations." Among other things, Moore argues that Foster fails to consider the very real potential for capitalist growth to be compromised by resource depletion and/or ecological collapse. In a similar vein, Davidson's work to develop the concept of the Effort Factor (Davidson, 2019; Davidson & Andrews, 2013; Davidson et al., 2014) brings those metabolic exchange processes, and their inevitably historical character, to the fore. The Effort Factor refers to the amount of effort, including labor and technology, required to convert raw materials into the stuff that people use. This effort describes an evolutionary process, defined by changes over time in the human and technological capacities utilized, and the changing quality of the raw materials themselves, which tends to decline over time, inevitably in the case of non-renewable materials like minerals and fossil fuels; historically in the case of renewables like soil (agriculture), fish and forests. This decline in the quality of raw materials requires a requisite increase in the effort required to convert those materials into things people consume, in turn accelerating ecological impact. This is perhaps most vividly observed in the extraction of oil. While new deposits continue to be discovered, in the aggregate the quality of deposits has been in steady decline over the past century, such that much of what remains is in deep undersea deposits or locked in solid substrates like bitumen and shale. This declining quality has compelled reliance on more extreme and costly methods of extraction, including deep-sea drilling, hydraulic fracturing, and oilsands mining. No manner of technological, market, or social "fix" will alter the fact that the costs of extraction increase as the quality of a raw material decreases.

Box 1: The End of Oil?
A just transformation is premised upon the phaseout of fossil fuels, but how might such a phaseout materialize? In previous years, many analysts spoke with confidence about "peak oil," a point at which the global volume of oil production will peak, followed by a precipitous decline, because the global reserves of oil will have become exhausted, leading to skyrocketing prices and economic turmoil. There were compelling reasons to support such an argument, given the exponential growth in production in previous decades, and the fact that all major deposits of oil had been discovered—there would be no new bonanzas on the horizon. The several estimated dates at which such a peak should have occurred have come and gone; however, and yet oil production continued to rise, due to a technological "bonanza," enabling access to deposits like bitumen and shale once considered too uneconomical to produce. The laws of metabolism nonetheless apply: these lower quality oil deposits require vastly larger amounts of effort to access and process. While the price at the pump has been kept low, the environmental impact has only continued to rise. How might the production of oil cease, when valuable reserves remain in the ground? To answer that question, we must begin to consider the *social* metabolism of oil, or more accurately, the goods and services that oil provides. How can the social metabolism currently provided by oil be met in other ways? Might social metabolism in fact be enhanced, by eliminating the environmental impacts, and the concentrations in capital and wealth enabled by oil dependence?

Shifting Metabolic Conceptualizations of the Economic Sphere

As these critical engagements with metabolism suggest, capitalism (of all kinds) is the "elephant in the room" of our nature-society relations: as that set of drivers that pose the most formidable violations of society's metabolic relationship with the planet. Just transformations thus require critical interrogation of plausible pathways of transformation from within this deeply hegemonic and global economic system. Prevailing policies that seek to improve environmental conditions more often than not entail market mechanisms, such as carbon taxes, that seek to avoid precisely such transformational disruptions, in order to protect economic growth, representing what has been referred to as a "glass ceiling" of environmental state activity (Hausknost, 2020). However, we are today experiencing a broad (re)emergence of public and scholarly discourses seeking to establish alternative, ecologically embedded visions of economic planning that to varying degrees embody metabolic thinking, among which one of the most prominent policy prescriptions has been the Green New Deal (subsequently reversed by President Trump). Emerging discourses include, for example, doughnut economics (Raworth, 2018), post- and de-growth (respectively Jackson, 2021 and Hinkel et al., 2020), and sharing and circular economies, as well as attempts to place wellbeing rather than earnings at the center of the economy (see Fioramonti et al., 2022 for an overview). In the limited space we have available, we pick up on one of these threads, which has begun to be taken up by policymakers and businesses alike, particularly in Europe: the circular economy.

The idea of a circular economy represents an influential and innovative contribution to the conceptualization of sustainable living and of societal metabolism. According to Kovacic et al. (2020: 15) "[t]he term 'circular economy' started being used in the 1990s, drawing from two seminal works: Kenneth Boulding's (1966) economics of spaceship earth, and Stahel and Reday-Mulvey's (1976) closed-loop economy." Kovacic and colleagues (2020: 17) note that social metabolism was a building block of the CE notion. This concept was then popularized and institutionalized by a British solo yachtswoman turned entrepreneur, Ellen MacArthur, who set up her foundation in 2010.

Associated with her record-making career as a racer and adventurer, alone for long periods on the high seas where she was obliged to use resources as efficiently as possible, MacArthur imagined an organization of the economy in which everything possible was reused. This contrasted with "linear" models of production where materials were extracted or mined, converted into products for use, and then disposed of at the end of their usefulness. She and her sponsors set up the Ellen MacArthur Foundation to take this idea forward. Kovacic et al. (2020: 18) note that the "circular economy blends in the concepts of system and throughput. The idea of the circular economy is to make the economic system more circular, by reducing throughput and transforming waste into an input." MacArthur and her team devised the well-known "butterfly" diagrams which, from the Foundation's first report in 2013, illustrated the ideal that, after use, all items should be reused as high up the value chain as possible. Correspondingly, production should, wherever possible, involve the reuse of existing items.

Commercial sponsors and business interests added weight to this conceptualization, creating a policy term with broad credibility. It was embraced, for example, by the European Commission (2019) and by the UK and Scottish governments (in 2018 and 2016 respectively). Clearly, the introductory idea of the circular economy (getting away from linear patterns of sourcing, use and disposal, removing wastefulness and promoting reuse) has a broad appeal. As depicted by Korhonen, Honkasalo and Seppälä (2018a, 2018b: 38):

> Materials should first be recovered for reuse, refurbishment and repair, then for remanufacturing and only later for raw material utilization, which has been the main focus in traditional recycling. According to CE, combustion for energy should be the second to last option while landfill disposal is the last option. In this way, the product value chain and life cycle retain the highest possible value and quality as long as possible and is also as energy efficient as it can be.

At one level, it is relatively easy to turn this into a policy objective, since it is about promoting repair and setting up systems to encourage the use of materials for remanufacture as in the motor industry. Thus, governments and firms can try to ensure that mobile phones can be repaired—at least for the most common defects, such as screen damage and battery replacement. Companies can take steps such as registering the composition of automobile parts to know which elements of vehicle interiors can be used in the production of future models. Such measures keep products in use for longer, reduce the need for primary production and cut the amount of material going to landfill or incineration. In the UK, the Ministry of Justice has adopted this line of thinking in relation to prison management. Instead of replacing damaged TV sets, it has adopted a policy of repairing them within the prisons themselves, reducing waste and allowing inmates to learn skills in maintenance and electronics (Ministry of Justice, 2024).

There are, however, two conceptual questions we need to ask. The first is about the link between circularization practices in firms and institutions and the circularity of the economy as a whole. From a metabolic standpoint, it is questionable whether economies can indeed be circular if they are simultaneously committed to economic growth, which entails escalating consumption and waste, or even to accommodating the projected global population growth to around 10.5 billion before 2100. Moreover, firms will still have an incentive to sell new product lines; innovation and obsolescence have been closely linked for decades and it is unclear what will motivate manufacturers to favor repair over the sale of novel products. Production can readily become a great deal less linear; whether the global economy as a whole can be made fully circular is entirely unclear.

The second point pertains to the potential reusability of many material resources, and, even for those materials that are ostensibly recyclable, the decline in resource quality after repeated cycles of reuse. The leading conception of a circular economy focuses on items such as metals, which are persistent, recoverable and potentially reusable. Aluminum, for example, can be recovered from applications (in vehicle bodies for example) and recast or extruded into new products, time after time. However, there are limits, even with the most suitable metals such as aluminum, which is regarded as very highly recyclable. There are currently some uses for which recycled aluminum cannot be substituted for primary-production metal because of requirements for purity and crystal size in certain applications. Other inorganic materials are capable of fewer repeated uses. And, of course, substances that we use to produce energy (things we burn or consume in a fuel cell or—indeed—eat) cannot be remade without applying at least the same amount of energy back into them as was released. This is essentially a fundamental energy, or metabolic, limit to circularity.

Within this same point, there are also complications relating to circularizing the bioeconomy. At one level, the biological realm is already highly circular. In natural habitats, biological resources are used and reconsumed down to the molecular level. But industrialized societies' attempts to circularize biological resources are generally less sophisticated. The by-products of the food and drinks industry, for example, are often fed to agricultural livestock (so much so that they are now commonly termed co-products). But such uses typically do not take advantage of the richness of biochemical ingredients in the by-products.

Aside from these conceptual issues with the range and reach of the circular economy, there are clearly political and practical questions about implementation. For example, it is notable that the Scottish Government's own document (2016) talks about moving to a "more circular" economy which seems to imply less specificity in the policy goal. In a similar way, Korhonen et al. (2018a, 2018b) suggest that the very concept may be "essentially contested"—that is, continually subject to discrepant interpretations much like "freedom" and "democracy" are. This implies that the concept may not facilitate the making of decisive assessments of policy options (see also Kovacic et al., 2020). Additionally, policy analysis needs to attend to the ways in which the circular economy links to other goals (notably to sustainable development and to decarbonization). On the face of it, these goals would appear to be in harmony. But differences in the ways the policy objectives are monitored will mean that at times the objectives can diverge: for example, some calculations of economic circularity pay less attention to sources of energy (see Haas et al., 2015) than would analysts of carbon neutrality and decarbonization.

As well as being adopted in policy circles, the term has proven attractive in business schools and among materials-minded engineers. For example, as well as promoting reuse and "upcycling" at the highest value, the desired shift toward circularity can also be conceptualized as the elimination of unnecessary material, sometimes referred to as "dematerialization." A commonly used example to illustrate this point comes from the analysis of suburban homes. During the day (and of course at night) there are frequently cars parked on driveways and in garages. These represent stocks of valuable materials that are being used in environmentally inefficient ways. An automobile may have a life of 10 or more years, but it will only be used for transportation for a fraction of that time, even though the metals in its body are effectively sequestered out of the circular economy for that entire period of time. The resources that are required to produce the metals and manufacture the cars have been expended and they are not available for other possible purposes. Power tools are another famous example; most households have them even though the tools are used on average for a few minutes a year. These and other underutilized household goods could be prime targets for sharing economy planning (e.g. Frenken & Schor, 2017).

This issue can also helpfully be thought of in terms of stocks and flows. The system of mass car ownership in the global north, China and oil-rich countries produces a vast stock of refined metal. One way to address this is to ask whether consumers actually want car ownership or whether what they want—and can afford—is public transportation services. During the COVID-19 pandemic Klaus Schwab (Executive Chair of the World Economic Forum, WEF) and Thierry Malleret published an analysis of the opportunities that could arise from a reassessment of market economies' leading economic model; they called this "The Great Reset" (2020). The book was accompanied by a very professional video presentation featuring key figures including King Charles III of the UK. One emphasis was on dematerialization—a widespread possible move away from ownership to the renting of services. An earlier document for the WEF—an imagined future outlined by Danish thinker and politician Ida Auken—was drawn into this discussion; that document had been given the deliberately provocative title of "I own nothing, have no privacy, and life has never been better." It spoke about the possibility of vastly reducing ownership by purchasing services instead of goods right across our lifestyle, and examined the possible rise of local economies where travel becomes unnecessary and thus much rarer. The WEF book and Auken's article became absorbed into conspiracy theories regarding alleged efforts to confiscate people's cars and houses, the abolition of ownership rights and limitations on citizens' freedoms (see BBC Monitoring and BBC Reality Check 2021).

The WEF stepped back from this vision to a degree, and events were overtaken by the 2022 Russian invasion of Ukraine and its profound economic consequences for the global economy. Nonetheless, Schwab and Malleret showed that the "paradigm" associated with dematerialization could have even more major implications than the desire to circularize our existing economy.

Metabolism and Collective Action

The concept of metabolism illuminates the necessity for reciprocal socio-ecological relations to support both human and ecological wellbeing in the long term. The concept has informed critical analyses of economic systems, as indicated in the previous sections. While less explicitly the focus of attention by metabolism scholars, ruptures in socio-ecological metabolism can provide the conditions favoring collective action. Therefore, we believe the concept of metabolism has much to offer to our inquiries of, and efforts toward, collective action—that most crucial of levers supporting Just Transformations.

The operationalization of Marx's writings has drawn attention to the strong potential for social mobilization in response to the disruptions posed by capitalism. This is the very premise of Marx's prognostications about a communist revolution emerging from capitalism. As noted by Baccini and Brunner (2023), while urbanization as an anthropogenic process has certainly posed metabolic challenges, cities are also sites of revolutionary advances in culture, knowledge and civil society which can and have been applied to the task of attending to our metabolic ruptures.

A metabolic lens can also inform inquiries into social mobilizations that emerge in response to the consequences of ruptures in socio-ecological reciprocity as well—the very social mobilizations that Martinez-Alier described as environmentalism of the poor. More recently, the escalation of climate anxiety, particularly among younger generations (Clayton, 2020) and the growing toll of climate-attributed extreme events constitute costs to our social metabolism, and enhancement of social mobilization potential. The experience of climate anxiety has been linked to activism, and indeed activism is seen as a valuable antidote to anxiety (e.g. Clayton & Parnes, 2025; Ogunbode et al., 2022; Schwartz et al., 2023). Similarly, direct experience with any of the many extreme events that are linked to global warming, such as fires, floods, and extreme heat, is linked to increases in climate concern and mitigation intentions among some groups (Ogunbode et al., 2019).

However, there is nothing in this narrative that implies in any deterministic sense that the materialization of metabolic rifts would necessarily result in social mobilization, nor can we presume that any potential transformational outcomes that emerge from such collective action would be just. While we can make the case that anything other than a Just Transformation is only likely to allow those metabolic rifts to persist, justice is nonetheless a quality that must be actively pursued, rather than presumed. Here too, perhaps paradoxically, the concept of metabolism has much to offer to inform our understanding of the conditions that are needed to support a vibrant and persistent social movement to support just transformation. A social movement, after all, can be conceived as a system, or "organism" that is seeking to flourish within a particular socio-political environment.

Personal engagement in collective action represents an enormous energy investment, with many potential sources of demotivation. Long-term engagement can lead to burnout for many actors when requisite rewards, and sources of rejuvenation are absent (Chen & Gorski, 2015). Meanwhile, as the environmental and climate crises continue to materialize at an escalating pace, the energy reserves of observers, as well as survivors, are compromised by the mental health impacts of those crises: anxiety can motivate action, but, without sufficient emotional support, it can also lead to withdrawal (Davidson & Kecinski, 2022). Research suggests that positive emotional rewards in the form of social belonging are at least as important as movement "wins" to ensure sustained engagement (Jasper, 2011). This in turn demands attention to justice and reconciliation. Metabolism also holds promise as a social movement frame: Akbulut and colleagues (2019) highlight the strong potential for an alliance between post-growth advocates and environmental justice movements, as both seek a politico-metabolic reconfiguration of our economies.

Conclusions and Discussion Questions

We have argued that the concept of metabolism aids environmental sociology by foregrounding the ways in which societies depend on their environments for materials, energy and a host of environmental services, such as cleaning and the breakdown of wastes. Metabolism directs our attention precisely to these exchanges with the natural systems that surround and suffuse human life and to the mechanisms through which these exchanges take place. Importantly, this approach facilitates a focus both on energy and matter.

Additionally, we have noted how metabolism can inform emergent framings that are particularly relevant to Just Transformations, including, first, a critical reexamination of that set of economic strategies embodied by the notion of circular economy, in the light of a metabolism lens. The idea of circularizing capitalist economies has become an influential policy goal and there are large-scale, imaginable advantages in terms of keeping materials in use and even substituting services for ownership of key consumption items. Still, attention to the overall societal metabolism encourages us to be skeptical of glib claims for circularity.

Second, we have explored the potential insights offered by applying the concept of metabolism to the emergence and perseverance of collective engagements to support Just Transformations. The concept of metabolism can serve as a discursive frame that can facilitate alliance-building, while also highlighting the need to attend to the sources of sustenance required for social movement mobilizations to ensure.

Discussion Questions

What are the potential scope and impact of dematerialization and the change-over from material goods ownership to receiving services, from transport and accommodation?

What are the possible metabolic limits to human societies' ability to circularize their economies?

What are the potential cultural sources of resistance and practical limitations to sharing economies?

Can we develop effective methods for assessing anthropogenic versus physiological metabolisms in differing socio-cultural contexts?

How can flows of materials and energy be accounted for in proposals for Just Transformations?

References

Akbulut, B., Demaria, F., Gerber, J. F., & Martínez-Alier, J. (2019). Who promotes sustainability? Five theses on the relationships between the degrowth and the environmental justice movements. *Ecological Economics, 165*, Article 106418. https://doi.org/10.1016/j.ecolecon.2019.106418

Auken, I. (2016, November). Welcome to 2030: I own nothing, have no privacy, and life has never been better. *World Economic Forum.* https://web.archive.org/web/20161125135500/https://www.weforum.org/agenda/2016/11/shopping-i-can-t-really-remember-what-that-is

Baccini, P., & Brunner, P. H. (2023). *Metabolism of the anthroposphere: Analysis, evaluation, design.* MIT Press.

BBC Monitoring, & BBC Reality Check. (2021, June 22). What is the Great Reset—and how did it get hijacked by conspiracy theories? *BBC News.* https://www.bbc.co.uk/news/blogs-trending-57532368

Boulding, K. E. (1966). The economics of the coming spaceship Earth. In H. Jarrett (Ed.), *Environmental quality in a growing economy* (pp. 3–14). Johns Hopkins University Press.

Bunker, S. G. (1985). *Underdeveloping the Amazon: Extraction, unequal exchange, and the failure of the modern state.* University of Chicago Press.

Bunker, S. G., & Ciccantell, P. S. (2005). *Globalization and the race for resources.* Johns Hopkins University Press.

Burkett, P., & Foster, J. B. (2006). Metabolism, energy, and entropy in Marx's critique of political economy: Beyond the Podolinsky myth. *Theory and Society, 35*(1), 109–156.

Capdevila-Cortada, M. (2019). Electrifying the Haber-Bosch. *Nature Catalysis, 2*(12), 1055. https://doi.org/10.1038/s41929-019-0404-2

Chase-Dunn, C. (1998). *Global formation: Structures of the world-economy* (2nd ed.). Rowman & Littlefield.

Chen, C. W., & Gorski, P. C. (2015). Burnout in social justice and human rights activists: Symptoms, causes and implications. *Journal of Human Rights Practice, 7*(3), 366–390. https://doi.org/10.1093/jhuman/huv011

Clark, B., & Foster, J. B. (2009). Ecological imperialism and the global metabolic rift: Unequal exchange and the guano/nitrates trade. *International Journal of Comparative Sociology, 50*(3–4), 311–334. https://doi.org/10.1177/0020715209105144

Clark, B., & York, R. (2005). Carbon metabolism: Global capitalism, climate change, and the biospheric rift. *Theory and Society, 34*(4), 391–428.

Clayton, S. (2020). Climate anxiety: Psychological responses to climate change. *Journal of Anxiety Disorders, 74*, Article 102263. https://doi.org/10.1016/j.janxdis.2020.102263

Clayton, S., & Parnes, M. F. (2025). Anxiety and activism in response to climate change. *Current Opinion in Psychology, 54*, Article 101996. https://doi.org/10.1016/j.copsyc.2024.101996

Cottrell, F. (2009). *Energy and society: The relation between energy, social change, and economic development* (Original work published 1955). AuthorHouse.

Davidson, D. J. (2019). The effort factor: An adjustment to our understanding of social-ecological metabolism in the era of peak oil. *Social Problems, 66*(1), 69–85. https://doi.org/10.1093/socpro/spy018

Davidson, D. J., & Andrews, J. (2013). Not all about consumption. *Science, 339*(6125), 1286–1287. https://doi.org/10.1126/science.1235599

Davidson, D. J., Andrews, J., & Pauly, D. (2014). The effort factor: Evaluating the increasing marginal impact of resource extraction over time. *Global Environmental Change, 25*, 63–68. https://doi.org/10.1016/j.gloenvcha.2014.01.001

Davidson, D. J., & Kecinski, M. (2022). Emotional pathways to climate change responses. *Wiley Interdisciplinary Reviews: Climate Change, 13*(2), Article e751. https://doi.org/10.1002/wcc.751

Department for Environment, Food & Rural Affairs, & Environment Agency. (2018). *Our waste, our resources: A strategy for England*. UK Government. https://assets.publishing.service.gov.uk/media/5c18f11740f0b60bbee0d827/resources-waste-strategy-dec-2018.pdf

European Academies' Science Advisory Council. (2015). *Circular economy: A commentary from the perspectives of the natural and social sciences*. https://www.easac.eu/fileadmin/PDF_s/reports_statements/EASAC_Circular_Economy_Web.pdf

European Commission. (2019, March 4). *Report from the Commission to the European Parliament, the Council, the European Economic and Social Committee and the Committee of the Regions on the implementation of the Circular Economy Action Plan* (COM(2019) 190 final). https://eur-lex.europa.eu/legal-content/EN/TXT/?uri=CELEX:52019DC0190

Ellen MacArthur Foundation. (2019). *Circular economy systems diagram*. https://ellenmacarthurfoundation.org/circular-economy-diagram

Lorenzo, F., Coscieme, L., Costanza, R., Kubiszewski, I., Trebeck, K., Wallis, S., Roberts, D., Mortensen, L. F., Pickett, K. E., Wilkinson, R., Ragnarsdottír, K. V., McGlade, J., Lovins, H., & De Vogli, R. (2022). Wellbeing economy: An effective paradigm to mainstream post-growth policies? *Ecological Economics, 192*, 107261. https://doi.org/10.1016/j.ecolecon.2021.107261

Fioramonti, L., Coscieme, L., Costanza, R., Kubiszewski, I., Trebeck, K., Wallis, S., ... & De Vogli, R. (2022). Wellbeing economy: An effective paradigm to mainstream post-growth policies? *Ecological Economics, 192*, 107261.

Fischer-Kowalski, M., & Weisz, H. (2016). The archipelago of social ecology and the island of the Vienna School. In H. Haberl, M. Fischer-Kowalski, F. Krausmann, & V. Winiwarter (Eds.), *Social ecology: Society–nature relations across time and space* (pp. 3–28). Springer.

Foster, J. B. (1999). Marx's theory of metabolic rift: Classical foundations for environmental sociology. *American Journal of Sociology, 105*(2), 366–405. https://doi.org/10.1086/210315

Foster, J. B. (2000). *Marx's ecology: Materialism and nature*. Monthly Review Press.

Frenken, K., & Schor, J. (2017). Putting the sharing economy into perspective. *Environmental Innovation and Societal Transitions, 23*, 3–10. https://doi.org/10.1016/j.eist.2017.01.003

Geissdoerfer, M., Savaget, P., Bocken, N. M. P., & Hultink, E. J. (2017). The circular economy—A new sustainability paradigm? *Journal of Cleaner Production, 143*, 757–768. https://doi.org/10.1016/j.jclepro.2016.12.048

Georgescu-Roegen, N. (1971). *The entropy law and the economic process*. Harvard University Press.

Giampietro, M., Mayumi, K., & Ramos-Martin, J. (2009). Multi-scale integrated analysis of societal and ecosystem metabolism (MuSIASEM): Theoretical concepts and basic rationale. *Energy, 34*(3), 313–322. https://doi.org/10.1016/j.energy.2008.07.020

Haas, W., Krausmann, F., Wiedenhofer, D., & Heinz, M. (2015). How circular is the global economy? An assessment of material flows, waste production, and recycling in the European Union and the world in 2005. *Journal of Industrial Ecology, 19*(5), 765–777. https://doi.org/10.1111/jiec.12244

Haberl, H., Wiedenhofer, D., Pauliuk, S., Krausmann, F., Müller, D. B., & Fischer-Kowalski, M. (2019). Contributions of sociometabolic research to sustainability science. *Nature Sustainability, 2*(3), 173–184. https://doi.org/10.1038/s41893-019-0225-2

Hausknost, D. (2020). The environmental state and the glass ceiling of transformation. *Environmental Politics, 29*(1), 17–37. https://doi.org/10.1080/09644016.2019.1625145

Hinkel, J., Mangalagiu, D., Bisaro, A., & Tàbara, J. D. (2020). Transformative narratives for climate action. *Climatic Change, 160*, 495–506. https://doi.org/10.1007/s10584-020-02761-y

Jackson, T. (2021). *Post-growth: Life after capitalism*. Polity Press.

Jasper, J. M. (2011). Emotions and social movements: Twenty years of theory and research. *Annual Review of Sociology, 37*(1), 285–303. https://doi.org/10.1146/annurev-soc-081309-150015

Jorgenson, A. K., & Clark, B. (2012). Are the economy and the environment decoupling? A comparative international study, 1960–2005. *American Journal of Sociology, 118*(1), 1–44. https://doi.org/10.1086/665990

Korhonen, J., Honkasalo, A., & Seppälä, J. (2018a). Circular economy: The concept and its limitations. *Ecological Economics, 143*, 37–46. https://doi.org/10.1016/j.ecolecon.2017.06.041

Korhonen, J., Nuur, C., Feldmann, A., & Birkie, S. E. (2018b). Circular economy as an essentially contested concept. *Journal of Cleaner Production, 175*, 544–552. https://doi.org/10.1016/j.jclepro.2017.12.111

Kovacic, Z., Strand, R., & Völker, T. (2020). *The circular economy in Europe: Critical perspectives on policies and imaginaries*. Routledge.

Martínez-Alier, J. (2002). *The environmentalism of the poor: A study of ecological conflicts and valuation*. Edward Elgar.

Marx, K. (1973). *Grundrisse: Foundations of the critique of political economy (rough draft)* (M. Nicolaus, Trans.). Vintage. (Original work written 1857–1858)

Marx, K. (1976). *Capital: A critique of political economy* (Vol. 1). Vintage. (Original work published 1867)

Ministry of Justice. (2024). *Circular economy strategy summary*. UK Government. https://www.gov.uk/government/publications/circular-economy-strategy-summary-moj/circular-economy-strategy-summary

Moore, J. W. (2011). Transcending the metabolic rift: A theory of crises in the capitalist world-ecology. *The Journal of Peasant Studies, 38*(1), 1–46. https://doi.org/10.1080/03066150.2010.538579

Moore, J. W. (2017). Metabolic rift or metabolic shift? Dialectics, nature, and the world-historical method. *Theory and Society, 46*, 285–318. https://doi.org/10.1007/s11186-017-9290-0

Nobre, G. C., & Tavares, E. (2021). The quest for a circular economy final definition: A scientific perspective. *Journal of Cleaner Production, 314*, Article 127973. https://doi.org/10.1016/j.jclepro.2021.127973

O'Connor, J. (1988). Capitalism, nature, socialism: An introduction. *Capitalism Nature Socialism, 1*(1), 11–38.

Odum, E. P. (1969). The strategy of ecosystem development. *Science, 164*(3877), 262–270. https://doi.org/10.1126/science.164.3877.262

Odum, H. T. (1995). Self-organization and maximum empower. In C. A. S. Hall (Ed.), *Maximum power: The ideas and applications of H. T. Odum* (pp. 311–330). University Press of Colorado.

Odum, H. T. (1996). *Environmental accounting: Emergy and environmental decision making*. John Wiley & Sons.

Odum, H. T. (2007). *Environment, power, and society for the twenty-first century: The hierarchy of energy*. Columbia University Press.

Odum, H. T., & Odum, E. C. (2001). *A prosperous way down*. University Press of Colorado.

Ogunbode, C. A., Doran, R., Hanss, D., Ojala, M., Salmela-Aro, K., van den Broek, K. L., & Karasu, M. et al. (2022). Climate anxiety, wellbeing and pro-environmental action: Correlates of negative emotional responses to climate change in 32 countries. *Journal of Environmental Psychology, 84*, 101887. https://doi.org/10.1016/j.jenvp.2022.101887

Ogunbode, C. A., Böhm, G., Capstick, S. B., Demski, C., Spence, A., & Tausch, N. (2019). The resilience paradox: Flooding experience, coping and climate change mitigation intentions. *Climate Policy, 19*(6), 703–715. https://doi.org/10.1080/14693062.2019.1639496

Park, R. E., Burgess, E. W., & McKenzie, R. D. (1925). *The city*. University of Chicago Press.

Polanyi, K. (1957). *The great transformation: The political and economic origins of our time*. Beacon Press. (Original work published 1944)

Ramos-Martin, J., Giampietro, M., & Mayumi, K. (2007). On China's exosomatic energy metabolism: An application of multi-scale integrated analysis of societal metabolism (MuSIASEM). *Ecological Economics, 63*(1), 174–191. https://doi.org/10.1016/j.ecolecon.2006.10.015

Ramos-Martín, J., Canellas-Bolta, S., Giampietro, M., & Gamboa, G. (2009). Catalonia's energy metabolism: Using the MuSIASEM approach at different scales. *Energy Policy, 37*(11), 4658–4671. https://doi.org/10.1016/j.enpol.2009.06.026

Raworth, K. (2018). *Doughnut economics: Seven ways to think like a 21st-century economist*. Chelsea Green Publishing.

Reuters. (2021). Fact check: The World Economic Forum does not have a stated goal to have people own nothing by 2030. *Reuters*. https://www.reuters.com/article/uk-factcheck-wef-idUSKBN2AP2T0/

Rice, J. (2007). Ecological unequal exchange: Consumption, equity, and unsustainable structural relationships within the global economy. *International Journal of Comparative Sociology, 48*(1), 43–72. https://doi.org/10.1177/0020715207072159

Schwab, K., & Malleret, T. (2020). *COVID-19: The great reset*. Forum Publishing.

Schwartz, S. E., Benoit, L., Clayton, S., Parnes, M. F., Swenson, L., & Lowe, S. R. (2023). Climate change anxiety and mental health: Environmental activism as buffer. *Current Psychology, 42*(20), 16708–16721. https://doi.org/10.1007/s12144-021-01414-8

Scottish Government. (2016). *Making things last: A circular economy strategy for Scotland*. https://www.gov.scot/publications/making-things-last/

Saito, K. (2023). *Marx in the Anthropocene: Towards the idea of degrowth communism*. Cambridge University Press.

Stahel, W. R., & Reday-Mulvey, G. (1976). *Jobs for tomorrow: The potential for substituting manpower for energy*. European Commission.

White, L. (1943). Energy and the evolution of culture. *American Anthropologist, 45*(3), 335–356. https://doi.org/10.1525/aa.1943.45.3.02a00020

Wolman, A. (1965). The metabolism of cities. *Scientific American, 213*(3), 179–190. https://doi.org/10.1038/scientificamerican0965-178

Yearley, S. (1992). Environmental challenges. In S. Hall, D. Held, & T. McGrew (Eds.), *Modernity and its futures* (pp. 117–167). Polity Press.

Yearley, S. (2023). Non-governmental organizations and the environmental movement: Challenges in climate change framing. In Z. Baker, T. Law, M. Vardy, & S. Zehr (Eds.), *Climate, science and society: A primer* (pp. 77–85). Routledge.

7

Demand

Sara Karimzadeh and Henrike Rau

Introduction

Growing demand for products and services, and the natural resources required to produce and provide them, has been a key contributor to multiple socio-environmental crises, most notably climate change. For example, the IPCC's report titled *Demand, services and social aspects of mitigation* (2022) asks if 'demand reduction is compatible with growth of human well-being?' In this report, the authors explore this question in great detail, suggesting that not only is it possible to maintain well-being, but also to enhance it while working to reduce global energy consumption (Creutzig et al., 2022). The main argument is that reducing demand for fossil fuels and resource-intensive products through interventions such as transforming consumption patterns, enhancing energy and material efficiency, and redefining service provision can significantly complement supply-side solutions. More generally speaking, there is growing recognition that a just transformation of society-nature relations needs to incorporate research-led efforts to reduce demand for energy and materials, especially by affluent nations, while recognizing variations in consumption over time and across societies. In particular, effective climate

S. Karimzadeh (✉)
Örebro University, Örebro, Sweden
e-mail: sara.karimzadeh@oru.se

H. Rau
Ludwig-Maximilians-Universität München, Munich, Germany
e-mail: henrike.rau@lmu.de

D. J. Davidson et al. (eds.), *Environment and Society*,
https://doi.org/10.1007/978-3-032-27059-7_7

change mitigation requires a concerted effort by affluent nations to reduce their demand for energy and material consumption.

To date, growth scenarios including Green Growth have been implicated in the persistent failure to adequately address climate change, necessitating calls for a reduction of demand in absolute terms (Hickel & Kallis, 2020; SRU, 2024). Given the ongoing failure to decouple global economic growth from resource consumption and carbon emissions, alternative approaches like Circular Economy, Consumption Corridors, and Sufficiency have gained traction. These strategies prioritize reducing energy and resource demand by pairing varying degrees of consumption reduction with improved well-being, meeting needs without overconsumption. Rethinking demand lies at the heart of these efforts while simultaneously advocating for enhancing well-being. In addition, studies rooted in practice theory have shed new light on the role of consumption (and, by extension, demand). These social-scientific approaches have called for a much broader interpretation of 'demand', focusing instead on consumption with its material elements, meanings and issues of competence.

Demand reduction strategies work in three ways: *avoid-shift-improve*. *Avoid* refers to adopting a minimalist lifestyle by consuming less, emphasizing a more or less radical change in socio-cultural norms; *shift* implies moving from one form of demand to another, often with a main focus on infrastructure (e.g. moving from energy-intensive air travel to high-speed rail for medium-distance travel); and *improve* refers to greening consumption by increasing efficiencies through technology (e.g. retrofitting buildings with better insulation and energy-efficient heating systems) (see Jarre et al., 2024). According to the A-S-I model, human agency, social practices and institutional dynamics should be brought in line to lower demand and control emissions, pointing toward an underlying consensus-oriented view of these layers as mutually complementary. Overall, it is believed that addressing demand offers a dual advantage: lowering overall resource consumption while improving human well-being. According to the IPCC report (2022), an explicit focusing on demand for services can broaden mitigation options that support low-carbon transitions while maintaining or improving lead to a more equitable distribution of resources and services, enabling low-carbon transitions without sacrificing quality of life.

Yet mainstream environmental sociology has been slow to address questions of demand, focusing instead on supply-side dynamics such as production systems and resource extraction. Exceptions include social-scientific and interdisciplinary research on (un)sustainable consumption that incorporates sociological approaches. In this context, a host of relevant perspectives have emerged on how to redefine and shift demand, including consumption corridors, circularity, sufficiency and degrowth approaches. Importantly, the role of routine everyday practices has been recognized as a key influence on creating new forms of consumption, with proposals rooted in practice theory to recraft or substitute resource-intensive (bundles of) practices. However, the attractiveness of such proposals varies considerably within and across countries, reflecting local and national consumption cultures, that is, particular place-specific ways of viewing consumption, its consequences for society and the environment and its linkages with everyday life. These consumption cultures shape demand by encouraging or discouraging the adoption of resource-intensive practices.

Still, to reconfigure demand toward more sustainable practices to ensure basic human needs are equitably met, further social-scientific knowledge is needed about the root causes of demand. This chapter approaches the issue of demand from an environmental sociology perspective, reflecting an attempt to close some of the existing knowledge gaps. In doing so, we base our argument on the premise that demand is not merely a reflection of consumer preferences but also a pivotal driver of resource allocation and environmental impact that reflects deep-seated social and cultural conditions. Using Iran and Germany as constraining contexts to identify underlying ideas of demand-related stability, change and questions of fairness and well-being related to consumption, we explore how the current approaches to study consumption have (not) addressed the issue of culture and cultural difference when discussing ways to reduce demand/consumption. We do so by examining already existing practices characterized by levels of consumption that meet both social and environmental sustainability goals and to demonstrate how different consumption cultures can help place demand within sustainable boundaries that promote social justice while safeguarding the planet. Here it will be shown that many of these alternative approaches are limited in terms of their sensitivity to cultural context, leaving significant gaps to be addressed in future social-scientific consumption research.

Box 1. Demand and Inequality

The question of how to reduce demand in a fair and equitable way has been central to critical social-scientific consumption research and socio-climate studies. Perhaps more importantly, it has also been the subject of public debate in countries struggling with high levels of social inequality (in addition to global and local environmental problems). Viewing demand as a fundamental aspect of social and economic life implies recognizing diverse implications for, and responsibilities toward 'others' when attempting to lower the material throughput of a given society for urgent ecological reasons.

Defining 'Demand': From Classical Economics to Culture-Sensitive Sociological Understandings

The concept of demand originates from the Latin verb *demandare*, composed of *de* (down or away) and *mandare* (to entrust or command). In its original sense, the word conveyed the idea of directing or instructing something to be done, highlighting an authoritative request or command. Over time, this evolved into its economic meaning, where demand refers to the desire and willingness to obtain a good or service in exchange for payment (Smith, 1951). In classical economics, demand is shaped by the interaction of price, consumer preferences, and income, all of which are believed to be influenced by rational behavior and the pursuit of utility maximization (Mankiw & Taylor, 2020). While demand, according to classical economics is understood within the self-regulating free market that maintains equilibrium without requiring government intervention, subsequent developments such as the Keynesian emphasis on the role of consumer and government intervention (Keynes, 1936) served to highlight the inherently relational nature of demand. What is evident in classical economics is that demand is fundamentally a market-driven phenomenon to meet individual preferences and interest. Today, economic definitions of demand—primarily referring to consumer desire or a need for goods and services—are widespread. The development of economic theories and the rise of industrialization have led to the exploitation of demand, with profound consequences for both nature and society.

Individualistic and economistic understandings of demand have also characterized some strands of social science work on the subject, especially during the second half of the twentieth century, a fact that has been criticized by prominent SC scholars (e.g. Miller, 2012, Warde, 2015). As Warde (2015) observes, '[n]arrow definitions [of consumption], for example, a "process of choosing goods" (Zukin & Smith Maguire, 2004, p. 173), tend to echo the formulations of economics and enthrone the consumer as the principal agent to be analyzed', thereby reinforcing 'the dominance of the neoclassical understanding of economic exchange' (p. 119).

However, these perspectives are gradually being challenged by a broader view of consumption as both the cause and the consequence of diverse forms of exchange in social life, including those involved in the (re-)production of routine everyday practices (e.g. Warde, 2005). Here, demand is seen as a fluid concept, shaped not only by individual interest and market dynamics but also by broader societal concerns. This creates space for integrating factors such as social and cultural influences, ethical considerations, emotions and concepts of care and responsibility for others. For instance, one might reflect on how demand for home heating is 'locked in' by existing infrastructures, systems of provision, and routine practices that reflect prevailing norms and prescriptions, making it rather resistant to change (Sahakian et al., 2020).

Agreeing with Warde (2005), we argue that the term 'demand', as it is commonly used, refers to a rather narrow understanding of material use (e.g. of energy, raw materials etc.), limiting its usefulness in the context of social-scientific inquiries into everyday life and its resource requirements. To address this limitation, we adopt a broader view of demand that incorporates (infra)structural conditions fueling consumption as well as socially negotiated and culturally/contextually rooted meanings that help to explain the persistence of resource-intensive and highly unsustainable practices. For example, we believe that the 'meaning of meat' (Biermann & Rau, 2020) can at least partly explain why people in highly developed countries like Germany consume unsustainable amounts of meat. In short, we define demand as consumption in a broader sense, with all its material, social and cultural aspects.

In what follows, we attend to three alternative social-scientific approaches that share our broad interpretation of demand: (a) demand in practice theory, (b) demand in circularity thinking, (c) demand in sufficiency studies. While highlighting the benefits of these alternative ways of thinking about demand, we also criticize their limited focus on cultural and social dimensions (section on consumption and culture). This helps to demonstrate the centrality of sustainability strategies and demand control measures that align and work with existing practices (rather than against them) and that recognize the essential role of consumption as a key material outcome of routine everyday practices.

Demand in Practice theory

An explicit focus on social practices and their consumption-related consequences has offered a credible conceptual alternative to hitherto dominant models of consumer behavior that focus (more or less) exclusively on attitudes, beliefs and choices (ABC) and that treat individuals as rational, atomized and largely self-interested actors, diverting attention away from the responsibility of business actors and political decision-makers regarding production-side and structural aspects.

Earlier calls by prominent practice theorists to 'move beyond the ABC' (Shove, 2010) have initiated a reconceptualization of consumption as the outcome of sets of oftentimes highly routinised, interlocking practices that are collectively adopted and carried by individuals. In other words, PT departs from dominant approaches emphasizing individual-level demand, offering an alternative that considers structural factors such as the material elements of practices, including infrastructure, building materials or components of appliances (Akenji, 2014; Shove, 2010). By extension, many scholars engaged in consumption research informed by PT question the high burden of responsibility that is often placed on individual consumers to 'manage' their demand, taking into account the latest evidence concerning the impact of their purchasing decision.

According to Braudel (1977a, 1977b) depicts everyday life as a field of what we might call practices: involve countless inherited actions that are repeated in daily routines and accumulate over time—becoming habits that both enable life and simultaneously organize and constrain us, our lives. Accepting such a definition, demand for energy and materials derives from people's engagement in everyday practices and their dynamics over time. For example, efforts to date to link the history of particular practices (e.g. cycling, freezing of food) to the material impact of individuals engaged in these practices have yielded fascinating insights into connections between structural opportunities and constraints and individual-level agency and its dynamics across the life course (Rau, 2024; Shove et al., 2009).

Calls have also been made to move away from a focus on individual practices, even if their reproduction appears to be particularly resource-intensive (e.g. home heating). Instead, scholars emphasize that practices regularly interact to form complex systems of interlocking practices that share a whole host of material elements (Spurling et al., 2013). In other words, consumption and the use of natural resources are integral parts of both individual practices and entire systems of interlocking practices. This is also reflected in Warde's (2005) statement that:

Consumption cannot be reduced to demand, requiring instead its examination as an integral part of most spheres of daily life (see Harvey et al., 2001). With this in mind, I understand consumption as a process whereby agents engage in appropriation and appreciation, whether for utilitarian, expressive or contemplative purposes, of goods, services, performances, information or ambience, whether purchased or not, over which the agent has some degree of discretion. In this view, consumption is not itself a practice but is, rather, '*a moment in almost every practice*' (p. 137, emphasis added).

The most recent expansion of practice theory, as articulated by Elizabeth Shove and others, focuses on reconfiguring demand to mitigate the adverse social and environmental effects of economic growth (e.g. Rinkinen & Shove, 2023). Reconfiguration, in this framework, occurs within and through practices. In their book *Conceptualizing demand*, Rinkinen et al. (2021) develop some key propositions to highlight the social side of demand creation: 'Demand is the outcome of social practice; demand is made, not simply met; demand is materially embedded and temporally unfolding; and demand can be steered via many forms of policy' (p. 84). According to this understanding of demand, it is not individuals who decide on their electricity use; rather, it is shaped by practices such as using home appliances or heating technology. Since practices are interconnected, a shift in one practice influences other practices and, consequently, overall demand. Put simply, the introduction of new routines or new appliances into our homes modifies existing practices, influencing both the nature and scale of demand.

Criticism of practice-theoretical approaches to consumption has tended to target the choice of unit of analysis (practices), instead calling for a more structural understanding of demand for material resources. Some argue that practices emerge from socio-technical regimes—such as market, culture, economy, and education—despite being carried out by individuals (cf. Hargreaves et al., 2011; Karimzadeh & Boström, 2022, 2024; Keller et al., 2022). As a result, demand cannot simply be seen as the result of social practices; rather, it is shaped by the interactions between practices, dominant regimes and the overall structure of society.

Considering the task of improving, shifting or avoiding demand, PT points toward the need to alter the material and social elements of resource-intensive practices, that is, everyday routines that require significant amounts of energy and materials. In the context of highly developed, industrialized countries, examples include meat eating, space and water heating and car-based mobility (Ivanova et al., 2020). Importantly, calling for a change in practices avoids the responsibilization of individual consumers, tackling the issue of an unfair and unequal distribution of the burdens of sustainability policy and action (e.g. financial costs, extra time needed). This said, individuals, through their role as 'carriers of practices' (Reckwitz, 2002) can play a significant role in reducing demand to sufficient levels, including through deliberate disengagement from resource-intensive practices.

Demand in Circularity

As stated previously, a traditional business-oriented perspective treats demand as a fluid concept that is continuously shaped by the current economic system, market relations as well as wider societal and regulatory conditions. As such, demand is largely driven by production systems that promote ownership and first-hand purchases. These production systems are designed to sustain demand by making ownership the default mode of consumption, reinforcing the idea that new is always better and discouraging alternative models such as renting, sharing, or buying second-hand.

In contrast, the circular economy (CE) perspective on demand focuses on how to persuade consumers to invest in reused and refurbished products, long-term usage, recycling, and repairing (cf. Kirchherr et al., 2017; Murray et al., 2017). Moving from linear traditional economic models to circular ones that focus on narrowing, slowing, or closing the resource cycle (Nußholz, 2017; Vidal-Ayuso et al., 2023), demand is conceptualized mainly as a shift from ownership to access (Bocken et al., 2016); a preference for services rather than products; increased interest in second-hand products; and reduced resource intensity. Moreover, concepts of circularity also envision the consumers as prosumers who are engaged in both the production/generation and use of goods, services or even essentials such as energy and who are expected to adopt practices such as repairing, sharing, and recycling (Camacho-Otero et al., 2018). A simple example can be closed-loop food systems where waste from one side (like the kitchen) can be input for another (compost for the garden).

Circularity relies on citizens' 'acceptance of practices' (Hobsons, 2016: 99, Greene et al., 2024). However, discussions of circularity often lack sufficient attention to the social dimensions of acceptance and adoption of practices (Geissdoerfer et al., 2017), including social and cultural influences such as norms, prescriptions and institutions and their rules on consumer behavior (see Camacho-Otero et al., 2018; Sahakian et al., 2020; Boyer et al., 2021; Vidal-Ayuso et al., 2023). For example, Corvellec et al. (2022) argue that the circular economy's narrow ideological underpinnings often overlook the socio-cultural dynamics that are needed to support a shift to full circularity. Similarly, various forms of bottom-up community initiatives in sustainable resource management that contribute to circularity, such as repair cafes, upcycling workshops or second-hand trading both online and through local flea markets and car boot sales, are regularly eclipsed by a focus on technological innovation. This also implies that the technocentric nature of dominant strands of the circular economy debate easily diverts attention away from addressing the socio-ecological challenges caused by overproduction and overconsumption (see Friant et al., 2021). This has led some scholars to advocate for the development and adoption of a broader concept of the 'circular society' (Jaeger-Erben et al., 2021). In practical terms, this would also help to better connect circular economy initiatives and the actual needs and practices of consumers.

Box 2. The circular economies of Iran

Beyond Western concepts of circularity: In societies like Iran, practices of circular economy are already embedded in the culture, as opposed to more abstract environmental concerns. Such practices often occur either due to traditional roots or financial constraints. For instance, repair services for electronic home appliances, as well as clothing and shoe mending, are widely available because they cater to a large market that sustains jobs but also are often more affordable than purchasing replacements. As a result, circular practices often emerge as the go-to option for many, helping individuals save money while revitalizing traditional professions and skills, and promoting a sense of job security for those involved. Although it cannot be considered a closed-loop circularity, it extends the lifespan of products and delays the need for new purchases. However, finding a reliable repair technician has become increasingly challenging due to the growing culture of replacing old items with new ones in an ever-evolving market, the difficulty of sourcing necessary repair parts, and the decline in technical skills. In line with this, Hunka et al. (2021) showed in their study on reused electronic products that trust and functionality can significantly reshape demand among individuals.

Further Discussion:

Reviewing such practices reveals that while the concept of a circular economy may not be widely recognized, its premises are often deeply embedded in daily traditions and practices in some communities. This underscores the importance of understanding local contexts and nurturing existing cultures of circularity to encourage demand patterns that align with planetary boundaries while promoting well-being. What are some traditions or practices in your own culture that promote circularity and help regulate consumption or demand? How might these be adapted to support sustainable development?

Sufficiency, Limits to Consumption and Demand

Recently, debates concerning the concept of sufficiency have also initiated a redefinition of demand, potentially curbing excess at both individual and societal levels. The sufficiency perspective challenges the conventional notion that 'more is better', instead advocating for a philosophy of 'living within limits' (Fuchs et al., 2021; Lage, 2022). Conceptualizing demand within the framework of sufficiency is a complex task since sufficiency itself, as a sustainability strategy, remains hitherto under-defined. As Hartmann (2024) notes, scholars in the field of sustainable consumption have yet to reach a consensus on whether sufficiency should be understood as 'enough', implying a broader perspective, or as 'strategies of reduction', reflecting a narrower interpretation. For us, sufficiency translates into absolute reductions in resource consumption. In sustainability literature, sufficiency is often viewed as an approach that centers on citizens' demand, along with the concerns, norms, and habits surrounding it. By promoting strategies like degrowth, it seeks to redefine well-being, shifting focus from consumption-driven metrics to more sustainable and holistic measures of quality of life.

Drawing on the ideas behind 'doughnut economics' developed by Kate Raworth (2018), the concept of 'consumption corridors' represents a variant of sufficiency thinking in relation to consumption. Here, socially negotiated limits to consumption that combine a social floor (lower limit, minimum) with an ecological ceiling (upper limit, maximum) intended to make it possible for people to live well without crossing key ecological thresholds (cf. Fuchs et al., 2021). The resulting corridor leaves sufficient room for individual-level agency to shape consumption across different domains of everyday life (Khan et al., 2023). 'Translating the concept [*of consumption corridors*] into policies would make it possible to induce fundamental changes in consumption without imposing specific lifestyles on individuals and without demonizing consumption' (Defila & Di Giulio, 2020: 315).

A core criticism of consumption corridors revolves around the difficulties of defining what people really 'need' (as opposed to 'want', for lack of a better word). Perhaps more importantly, the idea of curbing consumption/demand at the upper end tends to be met with considerable skepticism. This said, a small number of empirical inquiries into the acceptance of the concept of a consumption corridor shows that some aspects resonate with people while others do not. Work by Defila and Di Giulio (2020) on the acceptance of consumption corridors among Swiss citizens has provided a mixed picture. Using the semantic differential method, respondents were able to place themselves along seven bipolar scales that reflected key dimensions of the consumption corridor concept. The authors detect a general openness toward the regulation of demand through policy, with some controversy emerging around issues of scale (how much regulation), and by whom (the state). While some support exists within the population for efforts to curb excessive consumption at the upper end, there is less unanimity around negotiating the lower limit (Defila & Di Giulio, 2020). Similarly, survey-based research by Khan et al. (2023) on public support for five different eco-social policies in Sweden—three that prioritize an ecological ceiling (maximum income, wealth tax, meat tax) and two that focus on providing a social floor (basic income, working time reduction)—has produced interesting insights into what people consider to be acceptable. Depending on a number of demographic and attitudinal variables, support varies across these five policies, with working time reduction receiving the highest support. In contrast, a meat tax is viewed most negatively by survey respondents. In sum, finding common ground in Western liberal societies regarding an acceptable minimum and maximum of consumption remains an uphill struggle, with progress likely to be slow.

Box 3. Measuring well-being

The promotion of sufficiency and 'consumption corridors' often relies on generalized metrics of well-being that fail to capture the nuances of individual and cultural experiences. For instance, while sufficiency advocates may suggest limiting consumption as a pathway to well-being, individuals from cultures that prioritize abundance may resist sufficiency-oriented practices (Vighnesh et al., 2023). For instance, in Iranian culture, sufficiency used to be celebrated as a moral ideal in personal life. However, this ethos gives way to an equally powerful norm of generosity when others are involved: offering abundant hospitality or sharing one's resources is regarded as the highest form of care and social respect, even though these acts necessarily expand the total volume of consumption (Karimzadeh, 2025). It demonstrates that, in order to put sufficiency into practice, its applicability must be considered as narrowly as possible, while its scope should remain as broad as possible.

Further Discussion:

How do you feel about limiting your consumption choices? Consider this both as a voluntary decision, and a response to top-down consumption policies imposed on society.

Consumption and Culture

The role of culture in explaining consumption has often been mentioned. The most well-known and comprehensive sociological study articulating the interconnection between consumption/demand and culture is *Distinction: A Social Critique of the Judgment of Taste* written by Pierre Bourdieu in 1984.

In this book, the French sociologist inspired by Veblen's theory of *conspicuous consumption* (1899), masterfully argues that taste—the sociological driver of demand—is constructed within the social class, and the associated cultural capital influences what and how we consume and demand. Hence, from a sociological perspective demand is context dependent. This is not one-sided, though. Prominent trends might also rewrite cultural values and norms. Cultural commoditization, according to Appadurai (1986), refers to a social condition in which demand for certain products can transform cultural expression. For instance, the valuation of human-nature relationships may increase demand for environmentally friendly products.

However, given the oftentimes blurry boundaries of the term, it is not always clear what 'culture' actually refers to. Drawing on Anne Swidler's (1986) notion of culture as a toolkit of solutions to everyday problems, we believe that culture shapes and reflects the everyday practices of different social groups and, by extension, their use of a broad range of resources such as energy, water, land, or virgin and recycled raw materials. Similarly, differences between countries regarding the prevalence of resource-intensive practices such as meat eating, car-based mobility or home heating can be observed, reflecting culture-specific manifestations of everyday life and its organization. However, these differences have not yet been studied in any systematic way. Instead, rigorous cross-cultural comparative research such as Sonja Haustein's and Thomas Sick Nielsen's study of mobility cultures in 28 European countries remains the exception (Haustein & Nielsen, 2016).

Exploring evidence from different regions can offer valuable insights. The practice of consuming carefully and mindfully has deep roots in diverse cultures around the world, reflecting a rich and varied history. Rooted in different philosophical, religious, and cultural traditions, these practices often reflect a commitment to sustainability, spiritual growth, or communal well-being, responding to the challenges of resource scarcity or moral imperatives of moderation. For instance, the culture of mindful consumption used to be a very common culture among Iranians influenced by religious teachings, caring culture, experiencing periodic deprivation, and more recently the current socio-economic condition. For instance, squandering, especially of food and water, has long been viewed as undesirable among Iranians, as it reflects a lack of appreciation for the natural blessings one has been given. With their diverse yet fragile ecosystem, they have deeply ingrained the belief that a careful and sustainable approach to resource use is essential for survival. The irrigation system of *qanat* to protect, transport, and distribute water efficiently in hot and dry climate, in ancient Iran, is a great example that reflects an enduring awareness of environmental limits, adjusting demand, living well within the limits, but also preserving natural resources for future generations. This ecological consciousness has further strengthened the cultural emphasis on sufficiency and sustainability over time. Influenced by this, for Iranians, the concept of sufficiency revolves around seeking moderation in life and avoiding both excess and deprivation. This practice is also deeply rooted in their tradition, which views material possessions as an integral part of life that should be cared for. They focus on preserving their belongings to ensure they last longer. This approach aligns closely with the idea of 'true materialism', a concept introduced by Schor and Thompson in 2014, which has, in fact, been a common practice among Iranians for centuries. In that culture, caring for everyday materials reflects a commitment to protecting 'each other', as each item represents a significant investment of human labor in its production. Putting it simply, avoiding excessive consumption and rapid obsolescence is a common approach to control demand in Iran. However, this consumption culture should be primarily understood within the context of Eastern traditions, where conspicuous consumption is not a main priority. This culture is increasingly at risk of disappearing for various reasons, including the pervasive influence of Western consumption patterns. Additionally, embracing sufficiency is not appealing to everyone within the culture. Growing opposing perspectives links this way of life to misery, blaming it for diminishing quality of life, individual well-being, and happiness.

Knowing about such cultural conflicts, fostering a culture of responsibility and care—for people, resources, and the environment—is crucial for promoting sustainable consumption practices through sufficiency and enhancing the well-being of society as a whole (Karimzadeh & Boström, 2023). Consumption literature highlights mindful consumption practices (e.g. minimalism, sharing, repairing, etc.) for individuals, designed to thrive within a capitalist consumer society while protecting natural resources. However, for such sufficiency-oriented mindsets to have a lasting and meaningful impact, they must be deeply embedded in cultural values. Rather than introducing entirely new, unfamiliar behaviors, this shift requires drawing from existing cultural norms and practices that may benefit both sustainability and well-being.

Box 4. Practicing sufficiency: Some evidence from Iran

A brief look at history reveals that the practice of caring for material possessions and consuming thoughtfully was a common norm until the rise of capitalism in the West, which brought significant cultural and economic shifts. Prior to this, societies often emphasized the value of durability and repair. For instance, in many pre-industrial communities, clothing was meticulously mended, tools were sharpened rather than replaced, and household items were crafted to last for generations. In Iran, frittering away food or materials that could still be used was not only seen as a sin but also strongly condemned as a social norm. These behaviors reflect a collective ethos of sustainability and respect for the blessings of nature (Karimzadeh, 2025). A comparative study on ethical consumption in Sweden and Iran revealed that, in Western societies like Sweden, the market—facilitated by purchasing practices such as fair-trade products—is the primary arena for ethical consumption. In contrast, the Eastern version of ethical consumption, as seen in Iran, is mainly characterized by interpersonal, immediate activities such as caring practices (Karimzadeh & Boström, 2024). Here, collective agency is essential, as communities play a central role in reshaping demand to align with values of sufficiency and sustainability, prioritizing resource-sharing, social cohesion, and resilience over traditional consumption metrics.

Further Discussion:

Think about a society where resources are unequally distributed among the citizens and, as a result, well-being varies greatly across social classes. Do you think cultivating a 'culture of care' could effectively lower consumption and reduce 'unnecessary' demand? If so, what strategies could make this approach successful? If not, what are the primary obstacles limiting its effectiveness?

Box 5. Slowing down demand: Circular consumption in Germany

Culture also influences how people approach the issue of circularity, especially regarding Kirchherr's high-level Rs such as 'refuse' or 'rethink'. Taking Germany as an example of a highly developed, industrialized society, there is a noticeable lack of engagement in practices fostering circularity in areas such as mobility or housing. Proposals for car sharing, car shedding, or sharing living space appear to be met with considerable resistance, if not to say outright rejection. Instead, private ownership of living space, mobility tools or appliances appears to take precedence over any sharing arrangements. Deep-seated cultural values could be partially responsible for this reluctance to share things with others. Germans are generally quite risk averse, scoring high on the uncertainty-avoidance scale (Hofstede, 1984). This may be partly related to post-war experiences of extreme scarcity across Germany (and also shortages associated with the 'planned' economy in East Germany between 1949 and 1989). As a result, catering for all eventualities, including through the purchase of consumer goods that might be needed at a certain stage in the future, appears to be a significant driver of demand and consumption in Germany.

Discussion: The Future of Demand in a Climate-Constrained World

Engaging with demand—theoretically, empirically and in practical ways—is an essential precursor to planning for a Just Transformation, posing pressing questions regarding how much is consumed, but also whose needs are prioritized and how consumption patterns can be redirected toward equity and ecological sustainability. Our examination of core strands of social-scientific research that include the treatment of questions of demand/consumption has shown a broad range of definitions that reflect variations in how scholars interested in markets, practices, circularity and sufficiency view human needs, limits to demand/consumption, or the good life. This evidence of definitional diversity seems particularly important, given that rather narrow, economistic definitions of demand still dominate debates both within the academic world and beyond.

Embedded in wider discussions regarding the nature of sustainability transformations, ideas on how to reduce demand have certainly gained currency, with ways to avoid, shift, or improve the use of resources receiving different degrees of attention. To date, efforts to shift or improve have taken precedence over initiatives to avoid, thus risking perpetuating a cycle in which demand continues to grow, even if it becomes marginally more sustainable. We know by now that protecting our societies and planet is not merely a question of resource allocation or technological advancement; it is deeply embedded in cultural contexts and societal values. However, we notice a distinct lack of cultural sensitivity when it comes to social-scientific explanations of demand, leaving significant gaps in research, policy and practice.

Drawing on examples from Iran and Germany, we have shown that demand sits at the core of consumption cultures, which incorporate complex culture-specific norms and prescriptions related to concepts of needs and wants, freedom, fairness, work and care. People's actual performance of everyday practices, in turn, reflects the practical side of culture, representing tools for collectively solving everyday problems. Taking into consideration cultural heritage and sensitivity, and counting on local knowledge and practices regarding sustainability and resilience, are critical. Climate-related proposals and pathways with the aim of demand reduction must be grounded in local realities, capacities, and competences. Failing to recognize the significance of cultural variations in how demand is viewed and practiced inevitably leads to theoretical and conceptual gaps as well as weaknesses in the design and implementation of policies intended to reduce demand. The question is, to what extent do we, when developing sustainability ideas to protect our planet, take into consideration cultural nuances, capacities, and already established norms and traditions?

By critically engaging with demand—discussing its different framings in different sub-disciplines of environmental sociology, cultural underpinnings, and power dynamics—we lay the groundwork for transformative policies that not only shift or improve resource use, but also avoid unnecessary consumption altogether, ensuring a fair redistribution of material opportunities and a culture of sufficiency that is sensitive to local contexts and that fosters a Just Transformation.

Conclusion

Where to from here? Novel concepts of 'demand' that prioritize its links with social practices, culture and care offer a useful alternative to conventional economistic and individualistic approaches. Importantly, reconceptualizing demand also requires a new way of looking at transformation. Making so with existing practices and capacities (as opposed to calling for a radical shift in everyday life) as well as finding ways to utilize established cultural norms and prescriptions that encourage sufficiency might be a promising pathway. Aspects of Iranian and German consumption cultures presented in this chapter make it apparent that maintaining and potentially upscaling already existing, locally rooted practices that require limited resource input might be essential to any future transformation toward sustainability. In fact, recent environmental and energy policy controversies in Europe show that measures to reduce consumption only achieve legitimacy and acceptance by the public if they are deemed to be socially just, in addition to being ecologically sound.

References

Akenji, L. (2014). Consumer scapegoatism and limits to green consumerism. *Journal of Cleaner Production, 63*, 13–23. https://doi.org/10.1016/j.jclepro.2013.05.022

Appadurai, A. (Ed.). (1986). *The social life of things: Commodities in cultural perspective*. Cambridge University Press.

Braudel, F. (1977a). *Afterthoughts on material civilization and capitalism*. Johns Hopkins University Press.

Braudel, F. (1977b). *The Mediterranean and the Mediterranean world in the age of Philip II*. Harper & Row.

Biermann, G., & Rau, H. (2020). The meaning of meat: (Un)sustainable eating practices at home and out of home. *Appetite, 153*, Article 104730. https://doi.org/10.1016/j.appet.2020.104730

Bocken, N. M. P., de Pauw, I., Bakker, C., & van der Grinten, B. (2016). Product design and business model strategies for a circular economy. *Journal of Industrial and Production Engineering, 33*(5), 308–320. https://doi.org/10.1080/21681015.2016.1172124

Bourdieu, P. (1984). *Distinction: A social critique of the judgment of taste* (R. Nice, Trans.). Harvard University Press. (Original work published 1979)

Boyer, R. H. W., Hunka, A. D., & Whalen, K. A. (2021). Consumer demand for circular products: Identifying customer segments in the circular economy. *Sustainability, 13*(22), 12348. https://doi.org/10.3390/su132212348

Camacho-Otero, J., Boks, C., & Pettersen, I. N. (2018). Consumption in the circular economy: A literature review. *Sustainability, 10*(8), 2758. https://doi.org/10.3390/su10082758

Corvellec, H., Stowell, A., & Johansson, N. (2022). Critiques of the circular economy. *Journal of Industrial Ecology, 26*(2), 421–432. https://doi.org/10.1111/jiec.13187

Creutzig, F., Roy, J., Devine-Wright, P., Díaz-José, J., Geels, F. W., Grubler, A., Maïzi, N., Masanet, E., Mulugetta, Y., Onyige, C. D., Perkins, P. E., Sanches-Pereira, A., & Weber, E. U. (2022). Demand, services and social aspects of mitigation. In Intergovernmental Panel on Climate Change (Ed.), *Climate change 2022: Mitigation of climate change. Contribution of Working Group III to the Sixth Assessment Report of the Intergovernmental Panel on Climate Change* (pp. xx–xx). Cambridge University Press. https://doi.org/10.1017/9781009157926.007

Defila, R., & Di Giulio, A. (2020). The concept of "consumption corridors" meets society: How an idea for fundamental changes in consumption is received. *Journal of Consumer Policy, 43*, 315–344. https://doi.org/10.1007/s10603-020-09437-9

Di Giulio, A., & Fuchs, D. (2014). Sustainable consumption corridors: Concept, objections, and responses. *GAIA—Ecological Perspectives for Science and Society, 23*(3), 184–192. https://doi.org/10.14512/gaia.23.3.6

Friant, M. C., Vermeulen, W. J. V., & Salomone, R. (2021). Analysing European Union circular economy policies: Words versus actions. *Sustainable Production and Consumption, 27*, 337–353. https://doi.org/10.1016/j.spc.2020.11.001

Fuchs, D., Sahakian, M., Gumbert, T., Di Giulio, A., Maniates, M., Lorek, S., & Graf, A. (2021). Consumption corridors: Living a good life within sustainable limits. *Routledge*. https://doi.org/10.4324/9780367748746

Greene, M., Hobson, K., & Jaeger-Erben, M. (2024). Bringing the circular economy home: Insights from socio-technical perspectives on everyday consumption. *Cleaner and Responsible Consumption, 12*, Article 100157. https://doi.org/10.1016/j.clrc.2024.100157

Geissdoerfer, M., Savaget, P., Bocken, N. M. P., & Hultink, E. J. (2017). The circular economy—A new sustainability paradigm? *Journal of Cleaner Production, 143*, 757–768. https://doi.org/10.1016/j.jclepro.2016.12.048

Hargreaves, T., Haxeltine, A., Longhurst, N., & Seyfang, G. (2011). *Sustainability transitions from the bottom-up: Civil society, the multi-level perspective and practice theory* (Working Paper No. 2011–01). CSERGE.

Harvey, M., McMeekin, A., Randles, S., Southerton, D., Tether, B., & Warde, A. (2001). *Between demand and consumption: A framework for research* (CRIC Discussion Paper No. 40). University of Manchester.

Hartmann, E. (2024). Sufficiency as relations of enoughness. *Sustainable Development, 32*, 1–14. https://doi.org/10.1002/sd.3090

Haustein, S., & Nielsen, T. A. S. (2016). European mobility cultures: A survey-based cluster analysis across 28 European countries. *Journal of Transport Geography, 54*, 173–180. https://doi.org/10.1016/j.jtrangeo.2016.05.014

Hickel, J., & Kallis, G. (2020). Is green growth possible? *New Political Economy, 25*(4), 469–486. https://doi.org/10.1080/13563467.2019.1598964

Hobson, K. (2016). Closing the loop or squaring the circle? Locating generative spaces for the circular economy. *Progress in Human Geography, 40*(1), 88–104. https://doi.org/10.1177/0309132514566342

Hofstede, G. (1984). *Culture's consequences: International differences in work-related values* (Vol. 5). Sage.

Hunka, A. D., Linder, M., & Habibi, S. (2021). Determinants of consumer demand for circular economy products: A case for reuse and remanufacturing for sustainable development. *Business Strategy and the Environment, 30*(1), 535–550. https://doi.org/10.1002/bse.2636

Ivanova, D., Barrett, J., Wiedenhofer, D., Macura, B., Callaghan, M., & Creutzig, F. (2020). Quantifying the potential for climate change mitigation of consumption options. *Environmental Research Letters, 15*(9), Article 093001. https://doi.org/10.1088/1748-9326/ab8589

Jaeger-Erben, M., Jensen, C., Hofmann, F., & Zwiers, J. (2021). There is no sustainable circular economy without a circular society. *Resources, Conservation and Recycling, 168*, Article 105476. https://doi.org/10.1016/j.resconrec.2021.105476

Jarre, M., Noussan, M., & Campisi, E. (2024). Avoid–shift–improve: Are demand reduction strategies under-represented in current energy policies? *Energies, 17*(19), 4955. https://doi.org/10.3390/en17194955

Karimzadeh, S. (2025). Consuming with care: Insights into ethical consumption in Iran. *Frontiers in Sustainability, 6*, 1540113. https://doi.org/10.3389/frsus.2025.1540113

Karimzadeh, S., & Boström, M. (2022). Ethical consumption: Why should we understand it as a social practice within a multilevel framework? *Open Research Europe, 2*, 109. https://doi.org/10.12688/openreseurope.15069.2

Karimzadeh, S., & Boström, M. (2023). Ethical consumption in three stages: A focus on sufficiency and care. *Environmental Sociology, 10*(1), 1–11. https://doi.org/10.1080/23251042.2023.227797

Karimzadeh, S., & Boström, M. (2024). Cross-cultural perspectives on ethical consumption: A study of Swedish and Iranian citizens. *Journal of Consumer Culture*. Advance online publication. https://doi.org/10.1177/14695405241290920

Keller, M., Sahakian, M., & Hirt, L. F. (2022). Connecting the multi-level perspective and social practice approach for sustainable transitions. *Environmental Innovation and Societal Transitions, 44*, 14–28. https://doi.org/10.1016/j.eist.2022.05.002

Keynes, J. M. (1936). *The general theory of employment, interest and money*. Macmillan.

Khan, J., Emilsson, K., Fritz, M., Koch, M., Hildingsson, R., & Johansson, H. (2023). Ecological ceiling and social floor: Public support for eco-social policies in Sweden. *Sustainability Science, 18*(3), 1519–1532. https://doi.org/10.1007/s11625-023-01295-6

Kirchherr, J., Reike, D., & Hekkert, M. (2017). Conceptualizing the circular economy: An analysis of 114 definitions. *Resources, Conservation and Recycling, 127*, 221–232. https://doi.org/10.1016/j.resconrec.2017.09.005

Lage, J. (2022). Sufficiency and transformation: A semi-systematic literature review of notions of social change in different concepts of sufficiency. *Frontiers in Sustainability, 3*, Article 954660. https://doi.org/10.3389/frsus.2022.954660

Mankiw, N. G., & Taylor, M. P. (2020). *Economics*. Cengage Learning EMEA.

Miller, D. (2012). *Consumption and its consequences*. Polity.

Murray, A., Skene, K., & Haynes, K. (2017). The circular economy: An interdisciplinary exploration of the concept and application in a global context. *Journal of Business Ethics, 140*(3), 369–380. https://doi.org/10.1007/s10551-015-2693-2

Nußholz, J. L. K. (2017). Circular business models: Defining a concept and framing an emerging research field. *Sustainability, 9*(10), 1810. https://doi.org/10.3390/su9101810

Rau, H. (2024). Alltägliche Lebensführung und Nachhaltigkeit im Konsum. In *Handbuch Umweltsoziologie* (pp. 207–219). Springer Fachmedien Wiesbaden.

Raworth, K. (2018). *Doughnut economics: Seven ways to think like a 21st-century economist*. Chelsea Green Publishing.

Reckwitz, A. (2002). Toward a theory of social practices: A development in culturalist theorizing. *European Journal of Social Theory, 5*(2), 243–263. https://doi.org/10.1177/13684310222225432

Rinkinen, J., & Shove, E. (2023). Material culture and the circular economy. *Frontiers in Sustainability, 4*, 1158079. https://doi.org/10.3389/frsus.2023.1158079

Rinkinen, J., Shove, E., & Marsden, G. (2021). *Conceptualising demand: A distinctive approach to consumption and practice* (1st ed.). Routledge. https://doi.org/10.4324/9781003029113

Sachverständigenrat für Umweltfragen. (2024). Sufficiency as a "strategy of the enough": Curbing ecological crises and injustices. *GAIA—Ecological Perspectives for Science and Society, 33*(3), 275–281.

Sahakian, M., Rau, H., & Wallenborn, G. (2020). Making "sustainable consumption" matter: The indoor microclimate as contested cultural artifact. *Cultural Sociology, 14*(4), 417–437. https://doi.org/10.1177/1749975520944526

Schor, J., & Thompson, C. (Eds.). (2014). *Sustainable lifestyles and the quest for plenitude: Case studies of the new economy*. Yale University Press.

Shove, E. (2010). Beyond the ABC: Climate change policy and theories of social change. *Environment and Planning A, 42*(6), 1273–1285. https://doi.org/10.1068/a42282

Shove, E., Trentmann, F., & Wilk, R. (Eds.). (2009). *Time, consumption and everyday life: Practice, materiality and culture*. Berg.

Smith, V. E. (1951). The classicists' use of "demand." *Journal of Political Economy, 59*(3), 242–257. http://www.jstor.org/stable/1826437

Spurling, N., McMeekin, A., Shove, E., Southerton, D., & Welch, D. (2013). Interventions in practice: Re-framing policy approaches to consumer behaviour. *Sustainable Practices Research Group Report, 56*.

Swidler, A. (1986). Culture in action: Symbols and strategies. *American Sociological Review, 51*(2), 273–286. https://doi.org/10.2307/2095521

Veblen, T. (1899). *The theory of the leisure class: An economic study of institutions*. Macmillan.

Vidal-Ayuso, F., Akhmedova, A., & Jaca, C. (2023). The circular economy and consumer behaviour: Literature review and research directions. *Journal of Cleaner Production, 418*, Article 137824. https://doi.org/10.1016/j.jclepro.2023.137824

Vighnesh, N. V., Balachandra, P., Chandrashekar, D., & Sawang, S. (2023). How cultural values influence sustainable consumption behavior? An empirical investigation in a non-western context. *Sustainable Development, 31*(2), 990–1007. https://doi.org/10.1002/sd.2436

Warde, A. (2005). Consumption and theories of practice. *Journal of Consumer Culture, 5*(2), 131–153. https://doi.org/10.1177/1469540505053090

Warde, A. (2015). The sociology of consumption: Its recent development. *Annual Review of Sociology, 41*, 117–134. https://doi.org/10.1146/annurev-soc-071913-043208

Zukin, S., & Maguire, J. S. (2004). Consumers and consumption. *Annual Review of Sociology, 30*, 173–197. https://doi.org/10.1146/annurev.soc.30.012703.110553

8

Environmental Commons

C. S. A. Kris van Koppen, José Barrena, and Simon R. Bush

Introduction

Over the last few decades, "commons" have become a pivotal concept in both environmental social science and environmental action. Historically, commons referred to local natural resources, such as meadows, forests or waters that were managed and used collectively. At present, it refers to a much wider range of phenomena, including digital platforms and codes, scientific knowledge, and cultural productions. The concepts of commons and commoning have gained traction because they help rethink relations of property, democratic control, and use of material and immaterial resources. In this chapter we will apply commons and commoning to the environment—including natural resources and ecosystems—through the lens of spatial frames. The geographical size of environmental commons can vary from local or regional (e.g. common grazing grounds, or a lake or sea providing fish and other seafood) to global (e.g. the atmosphere with its ozone layer and greenhouse gases). Moreover, some local or regional phenomena, such as minerals in specific parts of the ocean floor or a unique species in a local ecosystem, can also be considered as global commons, especially when their existence and protection is seen as a common good of global importance. As these examples show, the designation, management and use of environmental commons are closely tied to issues of space and scale.

C. S. A. Kris van Koppen (Deceased) · J. Barrena · S. R. Bush (✉)
Wageningen University, Wageningen, Netherlands
e-mail: simon.bush@wur.nl

D. J. Davidson et al. (eds.), *Environment and Society*,
https://doi.org/10.1007/978-3-032-27059-7_8

The chapter proceeds as follows. We first provide more background to the concepts of commons, commoning, and space. Then, we analyse three major types of spatial framing. First, we investigate the relationship between territorial space and rights in nation-states, and the roles of state power vis-à-vis environmental commons. Second, we take up social-ecological systems as a key spatial frame put forward in natural resource management, with a strong link to commons, as is most famously manifest in the work of Elinor Ostrom. Third, we discuss flows and networks as a spatial frame that has emerged as an influential form of power in the world today. At the end of the chapter, we discuss the implications of our analysis, emphasizing the roles of all three frames in a just transformation that supports both the participation of local and historically marginalized groups in the governance and management of common resources, and the protection of our global environmental commons.

Commons, Commoning, and Space

Commons

Commons are increasingly considered a key concept in thinking about and working towards a socially just transformation to sustainability. In the domain of environmental protection and natural resource management, the debate on commons goes back to the 1960s. Most influential in this debate was the article *The tragedy of the commons* by the biologist Garret Hardin (1965). "Picture a pasture open to all", Hardin writes, "It is to be expected that each herdsman will try to keep as many cattle as possible on the commons" (p. 1244). According to Hardin, this will remain so, even when the pasture degrades from overgrazing, because the benefits of additional animals go to the individual owner, while the costs are shared by all. Therefore, it is economically rational for a farmer to add more cattle, even when it eventually destroys the commons: "Freedom in a commons brings ruin to all" (p. 1244). Hardin uses this story about commons to argue for governmental control of population growth, which he—and several other environmental authors at the time—regarded as a crucial threat to the environment. It should be added that his argument was seriously flawed, with sweeping claims without sound evidence and morally contentious advocacy for coercive control of the "freedom to breed" (p. 1246).

Despite these limitations, Hardin's argument has had broad and lasting appeal, for at least two main reasons. First, it employed the language of calculation and rational choice, which was taken up in economics to analyse natural resource management. Since Hardin, this field of environmental economics has developed further. Most notably, Elinor Ostrom's work has challenged Hardin's argument by demonstrating that many commons were, and still are, sustainably managed through specific institutional arrangements among communities and networks of users (Ostrom, 1990; Pellizzoni, 2018).

Second, the concept of commons has resonated with many environmental scholars and activists because it enables opposition to the commodification and privatization of environmental "goods". The concept also allows for a wider understanding of environmental resources—air, water, soil, and biodiversity—as the common heritage of present and future human and non-human generations. In other words, the environment is essentially a "common good" that needs collective action for its protection.

Commoning

It is in this sense of protecting and reclaiming the commons against privatization and commodification that the concept of "commoning" has emerged in literature (Hardt & Negri, 2008; Harvey, 2012). Harvey (2012) defines commoning as a practice:

> This practice produces or establishes a social relation with a common whose uses are either exclusive to a social group or partially or fully open to all and sundry. At the heart of the practice of commoning lies the principle that the relation between the social group and that aspect of the environment being treated as a common shall be both collective and non-commoditized – off-limits to the logic of market exchange and market valuations". (p. 73)

The concept of commoning expresses that the creation and persistence of a common requires continuing action of a collective. It may also imply an active struggle against processes of "decommoning", that is, dissolution of a common, e.g. into private property or into exclusive control by state authorities.

Many other authors have explored the concepts of commons and commoning along these lines (e.g. Bollier & Helfrich, 2014; Euler, 2018). In this strand of thinking, commons are defined as goods that are managed and used by a collective or community, in accordance with rules that are set by the collective and not dictated by market force or external power. Contra Hardin, these commons are not seen as sites of tragedy, but as promising arrangements for a just transformation of society.

Space

While it is not very hard to see how this can work for more traditional and established types of local commons, such as the common Alpine meadows of Swiss dairy farmers, application of the idea of commons to more large-scale and networked environmental goods raises several questions. In what ways do global market connections affect the functioning of commons? What is the role of national states in articulating commons? How should environmental commons be demarcated—and who decides on that? How can local actors set rules for a large-scale or even global common, and who is, effectively, member of this community?

Space is a fundamental dimension for understanding the biophysical environment, and the socio-political relationships that are involved in governing the environment. Spatial frames are conceptual lenses that can be used to understand and analyse (environmental) problems and to implement and justify solutions to these problems. In environmental policy and planning, spatial frames have taken a prominent role, as evidenced by concepts such as spatial policy, urban sprawl, land use planning, designation and zoning of protected areas, siting of industrial activities, dispersion and containment of pollution, and many others.

Traditionally, space has been framed as an exact and objective category. For instance, in the so-called Cartesian interpretation, space is measured in three dimensions of linear distance. This three-dimensional space can then be separated with clear boundaries, and nested according to well-defined levels of scale. In ecology, that could be organisms in their habitat, ecological communities, ecoregions, and biosphere. In government, it could be municipalities, provinces, and nation-states. However, as we aim to show in this chapter, space is a more complex phenomenon, among others, because technological innovations and processes of globalization have changed the role of spatial distances for social as well as ecological relationships. In many respects, these relationships have become networked across the earth, and it is the character of these networks, rather than linear distance, that determines which way and how fast information, money, and people move around the world.

In the light of this sociological understanding of space, we will now examine three major spatial frames in relationship to commons and commoning.

Territory and Rights Under Authority of the State

The Territorial Power of the Nation-State

The most prominent social framing of space is territory, which is typically defined as an area of land under the jurisdiction of a ruler or state (Oxford Dictionary). Spatial boundaries are as such central to the definition of any state territory. Over the last few centuries, nation-states have emerged as sovereign powers governing most of the world's terrestrial land. In the late twentieth century, this sovereign control was also extended offshore, granting nations authority over waters up to 200 nautical miles from their coastlines.

The concentration of territorial power with the rise of the modern nation-state is aptly characterized by Saskia Sassen (2006):

> Where in the past most territories were subject to multiple systems of rule, the national sovereign gains exclusive authority over a given territory and at the same time this territory is constructed as coterminous with that authority, in principle ensuring a similar dynamic in other nation-states. This in turn gives the sovereign the possibility of functioning as the exclusive grantor of rights. (p. 6)

As Sassen and other scholars have demonstrated, privatization and globalization challenge the role of the state in terms of territory, authority and rights. But the territorial power of nation-states remains substantial and often decisive in shaping the spatial organization of both resource extraction and environmental protection. Many multilateral environmental agreements explicitly reaffirm the authority of the sovereign states over their territories. This is demonstrated, for instance, in the preamble of the Convention on Biological Diversity which affirms "that States have sovereign rights over their own biological resources" (Secretariat of the Convention on Biological Diversity, 2005, p. 3).

As illustrated by the strong wording in the Convention on Biological Diversity, the territorial power of states is emphasized in the management of biological resources. In modern legal systems, natural resources within a nation's territory—such as minerals, oil or gas, but also wildlife and nature reserves—are considered the property of sovereign states. With this sovereignty, states maintain the right to control, extract, and tax their use (Moore, 2012). The distribution of rights and revenues of resource exploitation between the state administration and other actors—such as local-level administrations, Indigenous people, or private companies—serves as a strong indicator of a state's territorial power relative to other assemblages of territory, authority, and rights.

Spatial Planning and Designation of Nature Reserves

Over the last century, spatial planning has become a key governmental policy instrument for regulating tensions between (conflicting) economic, social, and political interests related to the allocation of environmental commons (e.g. Cullingworth, 2002).

Nature conservation is a case in point. The designation of areas based on their perceived "natural value" is a widely used policy instrument, determining how these spaces are viewed and managed. Classifying rural areas as nature–rather than agricultural, industrial or residential land–emphasizes their status as shared natural heritage while often downplaying the role of humans in shaping the land, and the customary rights of local and Indigenous people to access spaces and use resources. Often, this designation also reinforces state territorial control over these areas, albeit in different ways.

Chilean Patagonia is a good example of state control over peripheral spaces and peoples through the establishment of protected areas. Since the middle of the twentieth century, the Chilean state has reinforced sovereignty over its southernmost frontier by designating large national parks and reserves which have affected the rights and territorial organization of nomadic Indigenous people on both land and at sea. Unlike other cases where the establishment of protected areas has led to the forced eviction of Indigenous groups from their traditional territories, the establishment of large public protected areas in Chilean Southern Patagonia has largely disregarded Indigenous territories and mobile practices. This has in turn rendered Indigenous communities refugees in their own lands and waters (Harambour & Barrena Ruiz, 2019). On the one hand, the creation of protected areas by the Chilean state can be seen as an act of commoning, as these enclosures are intended to protect ecosystems and biodiversity for the benefit of national and global populations. On the other hand, however, the socio-cultural consequences—particularly the restrictions imposed on local and Indigenous communities to access and use resources—can be also considered decommoning (see Barrena et al., 2022, 2024).

Under conventional modes of protected areas designation, state authority is rarely disputed (Van Koppen, 2006). Instead, debates tend to focus on determining the most appropriate level of governmental control, whether at the federal, state, or provincial level. In many countries, there is a tendency towards decentralizing control to local levels of government. However, in the

European Union there is also upscaling of nature policy through, for example, the Bird and Habitat Directives, resulting in the designation of Natura 2000 sites in all member states (Evans, 2012). Similarly, the Biosphere Reserve Sites are designated under the intergovernmental Man and BiosphereProgramme by the Director-General of UNESCO as "learning places for sustainable development". Nevertheless, their nomination and jurisdiction remain under the authority of the states in which they are located (UNESCO, n.d.). While decentralization and centralization change how state control is exercised, they do not necessarily diminish state power. Ultimately, authority for nature protection continues to derive from state mandates or formal agreements, reinforcing its role as the primary governing entity in nature conservation.

Critiques of State-Territorial Power

Main strands of critique of state-led regulation of natural resources management focus on institutional fit and on decommoning effects of state power

Institutional fit, according to Young (2002), refers to the match between institutional arrangements and the defining features of the problems they address. Critiques of the state focusing on institutional fit stress the incapacities of the state to effectively regulate, implement, and monitor nature conservation, because of mismatches between environmental scale and jurisdictional level, as well as failures to adequately include non-state actors. As argued by Hooghe and Marks (2003), instead of state hierarchies (e.g. central state, provinces, and municipalities), function-specific networks crossing the state territorial levels may be better suited to overcome transboundary environmental problems. Examples are global city networks such as the Carbon Neutral Cities Alliance, and river basin partnerships like the Nile Basin Initiative. States may also lack the capacity to effectively manage nature protection at the subnational level (for an example, see BOX 1).

Box 1. Commoning by Local Groups

In Northern Patagonia, village-level organizations pushed the establishment of a Multiple-Use Coastal Marine Protected Area (AMCP-MU, its acronym in Spanish), which balances local economic activities with nature conservation. Although the AMCP-MU is a spatial tool for nature conservation by the state, local organizations led the process by defining the type of protected area suitable for their interests, networking with different territorial (fishers union, community board, local tourism organization, and Indigenous association) and extra-territorial actors (Environmental NGO, Regional University, Ministry of Environment), and, more recently, reaching an unprecedented agreement with the state for the co-managing of the area (see Gotteland et al., 2024). These organizations utilized state spatial instruments, such as AMCP-MU, to promote conservation in their territory in response to illegal fishing, resource overexploitation, and the expansion of polluting industries such as salmon farming.

A more fundamental critique focuses on the decommoning effects of state power over natural resources and nature protection areas. Brockington (2004), for example, demonstrates how effective state-led nature protection can also lead to the marginalization of local inhabitants. Studies on nature parks and reserves in Africa and Asia have similarly shown how the establishment of conservation areas cut off the access of rural communities to forest-related natural resources (e.g. Dupuis & Vandergeest, 1996; Wang, 2019). The example of Southern Patagonia in the previous section tells a similar story. Authors in this strand of critique contend that spatial planning, far from being merely a functional tool for organizing access and use to spaces and resources, can also function as a form of land and ocean grabbing (see for instance, Queffelec et al., 2021).

These critiques of state power, however, do not amount to an unequivocal rejection of the role and value of the state in the social organization of the space. While state power can be repressive, the lack of state control can lead to power abuse by other actors. Büscher and Dressler (2007), for example, show how a lack of adequate government planning in South Africa led to the rise of private game reserves and exclusion of local communities from once common lands. While numerous cases document how state power has forced or facilitated the annihilation of customary community rights to nature areas and resources—see for instance the analysis by Dell'Angelo et al. (2017) of large-scale land acquisitions—state authority can also play an instrumental role in safeguarding community rights. In this regard, Herrawan et al. (2022) conclude from research on social forestry in Indonesia "that the local commons cannot persist without the state's powerful protection of the current tenure and continued support" (p. 35).

Finally, in considering commons and the state, we observe that it is also possible—for specific cases or aspects of environmental protection—that the community whose rights are at stake is, legitimately, represented by the state. Examples are the designation of reserves for the protection of endemic red list species, or the establishment of protected areas with a national heritage value. As we underlined before, this would require that the state duly considers local rights on commons, and acts in democratic and inclusive ways. Our point is that in addition to local and global commons—which are central to commoning literature—there can be national commons.

Social-Ecological Systems and Commoning

Coupling Ecological and Social Systems

Critiques on state-level spatial planning, as described above, resonate well with advocates of social-ecological systems (SESs), a concept that has become an influential spatial frame for ecosystem management. Several streams of thinking have contributed to this development: adaptive co-management, which emphasizes self-organized processes of learning (Folke et al., 2005; Zuercher et al., 2022); the "ecosystem approach", developed under the framework of the Convention on Biological Diversity (UNEP, 2000); and most prominently, the work of Elinor Ostrom (2009) and colleagues (e.g. Cumming et al., 2020), who took up SESs as a flexible and inclusive framework for the design of institutions capable of governing environmental commons.

The system thinking that underpins SESs is outlined eloquently by Elinor Ostrom (2009):

> SESs are composed of multiple subsystems and internal variables within these subsystems at multiple levels analogous to organisms composed of organs, organs of tissues, tissues of cells, cells of proteins, etc. In a complex SES, subsystems such as a resource system (e.g., a coastal fishery), resource units (lobsters), users (fishers), and governance systems (organizations and rules that govern fishing on that coast) ... interact to produce outcomes at the SES level. (p. 419)

As Ostrom's description highlights, SESs offers a powerful spatial frame of coupled human and natural systems, structured through interacting levels of scale. Proponents of SESs (e.g. Chaffin et al., 2016) continue to argue that this integrative, dynamic, and multi-level approach provides a stronger foundation for designing resource management institutions than traditional nested hierarchies of state regulation.

SESs and Institutional Fit

Institutional fit is a key concept in the SESs approach. As it is conceptualized within the SESs approach, institutional fit can be summarized in three characteristics. First, the central importance of ecosystem dynamics in determining the optimal institutional scale. The institutional design should match the biophysical scale of vital biophysical processes. While nested hierarchies of scale remain important—similar to the conceptualization of territorial space—the primary focus is not on state authority structures but ecosystem processes.

Second, a preference to address problems at local levels of scale, as far as the ecosystem dynamics allow. This is epitomized in the aforementioned ecosystem approach, which states that "(m)anagement should be decentralized to the lowest appropriate level" (UNEP, 2000, p. 2). Shifting management to the lowest level is thought to increase stakeholder responsibility and participation, while also achieving greater effectiveness and social equity.

Third, a preference for local self-governance. This idea is, for example, apparent in Ostrom's (1990) design principles for governing common-pool resources. Principle 1 of her framework states that the boundaries of the resource and its legitimate users should be clearly defined, and principle 7 advocates the rights of local stakeholders to devise their own institutions, "not challenged by external governmental authorities" (Ostrom, 1990, p. 90).

Together these three characteristics of institutional fit in SESs hold two key assumptions. First, that the biophysical scale of the environmental problem, and its spatial boundaries, are identifiable. And second, that within these boundaries, institutions for natural resources management can be designed that will match the natural resource management challenge being confronted. These characteristics and assumptions around SES and institutional fit have provided an inspiring spatial framing that continues to appeal to scientists and policy makers alike. Institutional fit can be found, either explicitly or implicitly, in many forms of environmental management—including in community-based and collaborative management of nature parks, forests, and fisheries (e.g. Dixon et al., 2024). Perhaps the most salient example of SESs thinking is river basins. As seen in many countries around the world (e.g. Pahl-Wostl et al., 2021), efforts to reorganize water management have institutionalized in new and integrated types of river basin authorities (committees, platforms, agencies, boards) designated to the management of multiple water users in (sub)catchment areas.

Critiques of the SES Approach

Coupling and "fit" between ecological and social systems is highly complex in practice, and there are serious critiques on the applicability of SESs as a generic approach.

Managing ecological problems often undermines the ambitions for local control over local issues (Young, 2002). The benefits an ecosystem provides and the interests of stakeholders involved can vary from local scales (e.g. revenues from harvesting) to global scales (e.g. conservation of endemic biodiversity) (Hein et al., 2006). This means that different communities may have competing claims on the SES in question. Moreover, the socio-economic and biophysical drivers of ecosystem changes, as well as the institutions needed to cope with them, are more often than not characterized by market- and governance dynamics beyond the local or regional scale of the SES.

A consequential critique on mainstream applications of SESs is formulated by Eduardo Brondizio et al. (2009). Starting from an analysis of the Xingu Indigenous Park in the Amazon region, they argue that "no fixed spatial or temporal level is appropriate for governing ecosystems and their services effectively, efficiently, and equitably on a sustainable basis" (p. 259). Based on their analysis, they call for "a different style of scholarship from the dominant orientation of many social and physical scientists who study social-ecological systems" (p. 259), offering place-based management as an alternative—a concept we return to below in our discussion around networks and flows.

Another line of critique goes beyond the fixation of spatial or temporal boundaries, to instead focus on the systems approach underlying SES thinking. When applied to nature and natural resources, systems approaches tend to highlight instrumental values, or—when biodiversity protection is central—intrinsic values. The former considers nature valuable when it is useful for benefiting humans, while the latter considers nature valuable independently of human use or perception. Both types of values leave only limited room for the relational values of communities involved—those values built through the historical connections and multiple relationships between people and nature (cf. Himes, 2024). As Saunders (2014) argues, such deficits in systems thinking help to explain the success of commons with deep historical roots and the difficulties of establishing new commons in, for example, community-based resource management projects. Where common social learning histories related to resource use or conservation are missing, and the norms and practices of the community diverge from those assumed by the project developers, the process of commoning is largely undermined.

SESs, Commons, and Rights

Together these critiques show that in its simplified form the SESs approach often fails to deliver a just and effective resource management. Nevertheless, SESs offer a spatial framing that responds to critiques of the state and seeks to avoid scale mismatches. The SESs approach addresses institutional fit by developing self-organizing institutions for territorial management that align with the biophysical scale of environmental problems. Additionally, it invokes a renewed appreciation of "commons" by refuting the critique of Hardin and identifying conditions under which both traditional and new commons can be successfully managed and used by communities directly related to them. Efforts to strengthen environmental commoning can seek to address these critiques. One way is to refine existing institutional arrangements for just and effective natural resource management. For example, state institutions can be geared to giving more room for local institutions, through decentralization and participation, while continuing to protect local ecosystems from degradation and commodification (e.g. Gómez-Baggethun et al., 2013). Alternatively, in cases where non-state actors play a crucial role and cross-scale connectivity extends beyond the reach of a single state (or where states prove to be incapable of environmental reform or are, in fact, obstructing it) spatial frames that extend beyond the territorial power of states may offer new possibilities for governing environmental commons. This is what we turn to now.

Box 2. Justice and Effectiveness

Governance of commons, we argue in this chapter, should be just and effective. In Ostrom's analysis of common property resources, the emphasis is on effectiveness. The institutions governing a common are considered successful when they enable sustainable use of the common's resources over a long period of time (Ostrom, 1990). But what do we mean with just in regard to governing commons? In answering this complex question, we may think along three dimensions of environmental justice. Distributive justice implies that the benefits and costs of managing a common are fairly distributed among the community engaged in it. In determining what is fair, the efforts made by members, and the period of engagement with the common—which may go back far in history—should be taken into account. Procedural justice means that community members can meaningfully participate in decisions regarding the common. This is particularly important for members who are vulnerable and dependent on the common for their livelihood. Recognitional justice is about inclusion and respect for different values and sorts of knowledge. For many people engaged with nature, non-human beings need recognition as well, as part of the community. Just transformation in governing commons requires institutions that support these different dimensions of justice.

Globalization and the Space of Flows

Networks and Flows

In contrast to the scalar spatial frames of state territory and SESs and their nested geographical hierarchies, spatial frames focused on networks and flows take a different starting point. Rather than being divided in well-demarcated territories, pace is made up of places connected by various forms of relations that constitute social networks and various material and informational ideational flows (Castells, 2000). From this perspective, places are seen as the intersection of multiple flows of people, materials, knowledge, and technology. Likewise, boundaries are not seen as fixed lines in the space, but rather as the outcome of social relationships, and more specifically, the outcome of power relations (Bush et al., 2015).

Several authors have suggested concepts within the spatial frame of networks and flows. John Urry (2000) proposed concepts of *regions, networks and fluids* to theorize the mobilities that characterize our globalizing world. Bulkeley (2005) discusses *networks of spatial connectivity* to highlight the role of non-scalar geographies of space. Mol and Spaargaren (2006) put forward *governance of flows* as a promising way to understand environmental change. Despite this diversity, many of these concepts are based on the work of Castells (2000), who analyses the globalizing world with the concepts of *network society*, *space of places* and *space of flows*.

Castells describes the space of flows as a combination of three layers: (1) networks of electronic exchanges; (2) nodes and hubs that these electronic networks connect with and that program and switch the functions of the networks, including the global financial system; and (3) the spatial organization of dominant managerial elites.

Castell's space of flows, however, does not permeate the entirety of human experience. A majority of people perceive their space as place-based—with place understood as "a locale whose form, function, and meaning are self-contained within the boundaries of physical contiguity". (Castells, 2000, p. 453). A town, a city, or a quarter in a city, which is recognized by its residents as their common daily living environment, is a place in this sense.

Just as a city park, like New York's Central Park, can be a place, or rural areas, nature parks, and other kinds of landscapes.

Governing Commons in a Network Society

While biophysical flows play crucial roles in environmental governance, environmental protection and conservation of nature almost always involve the management of places. Environmental governance, therefore, is crucially concerned with the interactions between specific flows and the ways these interactions "co-produce" specific places. This goes for both global and local commons. There are innumerable examples of these processes, but we will focus on two: networked views on nature areas, and "territorialization" as a process of interaction between territories and socio-economic flows.

Nature Networks and Biodiversity as a Global Common

Over the last 25 years, networked views on nature have gained popularity. Some nature networks, like the National Ecological Network in the Netherlands, have actual biophysical connections in the form of green corridors. In other cases, such as the Natura 2000 network in the EU, or the UNESCO World Network of Biosphere Reserves, the connections are socio-political. These areas are not connected by physical continuity but by shared socio-political frameworks and conservation strategies.

A very influential example of a global, networked approach to nature conservation is the concept of *biodiversity hotspots*, introduced by Myers et al. (2000) and promoted by Conservation International. The 25 hotspots identified by Myers et al. are defined as "areas featuring exceptional concentrations of endemic species and experiencing exceptional loss of habitat" (p. 853). The hotspots are designated on the basis of biological properties, namely the number of endemic plant species and the loss of vegetation in the past—to qualify as a hotspot, it should have lost 70% or more of its primary vegetation (Myers et al., 2000).

Using Castells' framework, the space of flows in this case consists of global networks of conservation biologists, NGOs, funding agencies, and sponsors. The places are the designated hotspots, concrete geographical areas with local communities and institutions, but (re)defined from a specific global conservation logic, centred around the "calculus of biodiversity" (Vane-Wright, 1996, p. 332). From the perspective of commoning, these areas are framed as global commons, protected for the well-being of a global community, which many conservationists argue includes both human and non-human beings.

This networked logic has significant consequences for the places it touches (Van Koppen, 2006). Being labelled a "hotspot" alters how places are integrated into globally networked flows of money, scientific knowledge, and governance. The hotspot approach resembles the SESs approach in prioritizing conservation values and biophysical reasoning, and can be criticized for similar reasons. While the arguments for preserving biodiversity hotspots are compelling and often justified, conservation efforts must also be weighed against place-based values and institutional arguments to ensure effective and just ecosystem management.

Territorialization

Another example of how global networks and flows shape territories—or *maritories* in the case of marine environments—is provided by Barrena et al. (2024). They examine territorialization and counter-territorialization in the Beagle Channel, Chilean Southern Patagonia, where global networks tied to salmon farming and nature conservation establish marine enclosures that impact the livelihoods and mobility of the Yagán, a nomadic seafaring people. The authors define *territorialization* as a strategy of setting boundaries to control resources and populations. However, they also show how local actors, such as the Yagán, engage in *counter-territorialization*, resisting these enclosures by attempting to enrol globally articulated networks of scientists and NGOs to help reshape control over the discourses and regulatory processes to exclude salmon farming from their maritory. In this case, commoning is manifested in the establishment of the Cape Horn Biosphere Reserve, a five-million-hectare area encompassing both marine and terrestrial environments, administered by a public–private consortium of research institutions. While, on the one hand, this UNESCO site can be understood as a global common for sustainable development, on the other hand, Indigenous and local communities have argued that spatial claims to protected areas by global conservation networks are treated more favourably by the state than local spatial claims to land and sea by the Yagán people. Thus, in this case, it is possible to infer that global commoning is simultaneously producing local decommoning, although the physical boundaries of the Cape Horn Biosphere Reserve were also a relevant spatial tool used by the local counter-territorialization network against salmon farming.

Box 3. Global Commons and Local Access
Alignment of global and local commons is very well possible. Climate mitigation on a global scale is needed to halt degradation of local agricultural commons. Many Indigenous forms of collective nature management and use have contributed to global biodiversity conservation and continue to do so. But there are also conflicts. Particularly the discourse of biodiversity as a global common, supported by the symbolic and financial power of international nature conservation NGOs, is frequently used to restrict local access to nature without proper justification (Ribot and Peluso, 2003). A narrative of global commons, therefore, can also be used to legitimize local processes of decommoning. Thus, it is important not to take narratives of commons at face value. What is needed, particularly in cases of local commons, is careful empirical investigation of the different mechanisms involved in access to local commons (see Ribot and Peluso, 2003), as well as an explicit and consequential voice of local communities in the governance of these commons.

Linking Value Chains and Territories

While biodiversity hotspots and other global conservation networks represent an encounter between the global flows of conservation and places, a potentially even more influential dynamic is the interaction between production and consumption chains and places. Conventionally, value chains (also referred to as commodity chains) are seen as linking economic actors: consumers, firms, producers. However, as argued by Bush et al. (2015), this picture has shifted to include far broader networks of extra-transactional actors that connect environmental and economic practices across different places and environments.

Research on certification, risk management, supplier upgrading, and conservation in value chains has emphasized the role of collaboration between producers in specific areas and global buyers in sustainably managing the area's natural resources. This approach, variously labelled as a landscape or jurisdictional approach to sustainable sourcing is promoted through partnerships between private sector buyers and NGOs. The aim of these approaches, in the words of Kissinger et al. (2013) is to "provide a framework to deliberately work beyond the farm scale to support food production, ecosystem conservation, and rural livelihoods across entire landscapes in an integrated manner" (p. 1).

Similar approaches are emerging under the labels of area-based or place-based management (Bottema et al., 2021). Recognizing the interconnected and multi-level nature of resource management, they seek practical responses to fragmented authority and spatial and temporal mismatches in managing terrestrial and marine resources. Returning to Brondizio et al. (2009), these place-based approaches differ from prevailing SESs models because they view "places as complex and dynamic systems that are open rather than closed in character" (p. 271) and because they feature participation of government and other stakeholders at different levels.

By opening up to flows and networks, landscape or place-based approaches create theoretical and practical options for value chains to intersect and assert influence. For instance, Vandergeest et al. (2015) examine how value chains and territories interact in seafood certification. They show how certification can lead to the formation of "sustainable territories"—such as communities of shrimp farms in their coastal environment or designated fishing grounds (see also Bottema et al., 2021). This process, which they term "assembling sustainable territories", involves government agencies, value chain actors, and non-state organizations. As they demonstrate, this territorial assemblage can lead to "green grabbing"—akin to the notorious practices of land grabbing—while in other cases it fosters a more sustainable reconfiguration of value chains aligned with the place-based approach as advocated by Brondizio et al. (2009).

Non-scalar perspectives, commons, and rights

Commoning, from the perspective of flows and networks, points to the problematic nature of nested hierarchies of territorial scales when faced with networked social relations that extend through bounded territories. As we argued above, nation-states as well as communities in SESs cannot tackle problems with socio-economic and biophysical features that play out on a global scale if value chains, networks and flows are not considered. However, we also concur with Bulkeley (2005) in arguing that a polarization of the debate into "scalar" and "networked" perspectives should be avoided. Instead, hierarchies of geographical scale and networks of spatial connectivity can be better seen as mutually constitutive aspects of spatial framing.

Likewise, Castells pleads for building cultural, political and physical bridges between the space of places and the space of flows—both in theory and practice.

Following this line of thinking we suggest that, in the context of global flows and networks, successful commoning requires the establishment of institutions that facilitate a socially and environmentally sustainable match between a specific place and the global flows and networks it is embedded in.

Discussion and Conclusions

We have presented a social interpretation of space and environmental commons as sets of material dimensions vested with social and institutional meanings. A key conclusion from our investigation is that processes of commoning (and decommoning), as well as our ability to understand them, are dependent on spatial frames. We examined three spatial frames characterized by the concepts of territory, social-ecological system, and networks and flows. We have connected these frames to commoning, understood as social practices that (re)produce a collective and non-commoditized relationship between a social group and an environmental common, whether at a local, national, or global scale. We suggest that each of the spatial frames can offer valuable insights for theorizing and practically designing pathways for just transformation, depending on the specific situation. At the same time, we argue that each spatial frame has strengths and weaknesses that need to be taken into account in determining their role in processes of commoning.

While still dominant in conventional environmental and spatial policies, state-based frames have clear limitations. Spatial frames guided by state territorial authority—such as those related to the creation of protected areas—have contributed to decommoning terrestrial and marine spaces, affecting customary rights of local and Indigenous people. This, in turn, could jeopardize efforts towards a just environmental transformation. Besides, it is important to highlight that in the current globalized world the authority of states and their associated institutional arrangements are continually shaped by global networks (Sassen, 2006). Despite these weaknesses, the state remains an indispensable power-container and regulator, which can aspire to keep or regain legitimacy through the just attribution of territory, authority and rights.

The social-ecological systems approach, particularly in the way it is elaborated by Ostrom and her collaborators, is in several respects akin to the protection and maintenance of local commons, while also incorporating, through its ecological orientation, aspects of global biodiversity commons. A weakness is the difficulty of setting boundaries to the impact of cross-scale flows and networks; this weakness is acknowledged by most advocates, and countered by concepts such as nested systems and telecoupling—that is, close connections through material and immaterial flows between distant human and natural systems (e.g. Hull & Liu, 2018).

Given the importance of global commons, and the impact of global social, economic and political processes on local commons, we argue that the network and flows frame provides a highly valuable approach for understanding how different types of groups (from private sector to NGOs to Indigenous groups) shape the commons—for better and worse. On the positive side, we see new forms of engagement between value chain and landscape-based approaches where ecosystem management and actions of producers and consumers are tied together in global production and consumption chains. It is worth noting here that commoning does not exclude selling goods produced from the common on the market—as for example, cheese from the alpine pasture commons in Switzerland. What matters is that the common is managed and used collectively, by and for the community (Harvey, 2012, p. 74). Moreover, as the Chilean cases exemplify, supranational civil society networks can support local communities in protecting and re-establishing their commons. On the negative side, as aptly observed by Dell'Angelo et al. (2017), common goods successfully managed by networks of local actors under inclusive forms of governance are particularly vulnerable to the "exogenous" forces of powerful market actors, often supported by government interventions, leading to a significant number of land grabbing processes affecting common goods.

Our final, overarching consideration is that commoning is a major component of working towards a just transformation in the environmental domain and that local, national, and global environmental commons are not necessarily excluding one another. Local commons can, in many respects, align with nationally and globally established commons. What is important is that the communities involved in commoning on these different levels of scale have a voice in these processes, through local participation, democratic procedures in the state, and global networks. This may sound abstract and idealistic, and in many respects it is. Nevertheless, there are concrete examples of deliberation processes within environmental governance that make such an attempt. On a global scale, we can think of the Rio Summit on Environment and Development in 1992, or the Conference of Parties in Paris, 2015, where the Paris Agreement on climate policy was crafted. On these meetings—and many similar ones—not only governments were represented, but also NGOs, in varying alliances, with many of them voicing local common interests as well as global ones. On a local scale, we can think of the counter-territorialization case in Southern Patagonia, described above. Such examples may inspire us to further develop practices of commoning towards a just transformation. Rather than contrasting the different spatial framings against each other, it is important to contrast them with the forces of decommoning, that is, the dynamics that force social, ecological and political interactions into the mould of power-based, transactional relationships under domination of economic and military power. Commoning, in this light, is what brings us together against the threat of a state- and market-based territorialization and commodification of environmental resources.

Discussion Topics and Questions:

If you think about nature and environment in the region you live in, what would you see as important commons? To which communities do these commons relate? Do you see threats of decommoning (for instance, by market forces) for these commons? And how could such threats be countered?

Institutional fit is a key aspect of commons governance. Considering this, what types of mismatches between environmental scales and jurisdictional levels can be identified in your region, and what approaches might help resolve them?

The chapter analyses the ambivalent roles of states with regard to commons. What do you think of the roles of the state? For a group discussion, you could choose a concrete case of a state and a specific common within this state, for example an endangered endemic species.

The chapter discusses the challenges of governing global environmental commons, such as the atmosphere with its ozone layer and greenhouse gases. What strategies do you think could help foster the participation of local actors in the governance of these kinds of commons?

A highly actual debate on commons concerns the mining of minerals on the ocean floor. If you were charged with organizing a global conference to make agreements on how to govern this common, who would you invite, and what would be your agenda? If you prefer, choose another case of global commons.

References

Barrena, J., Harambour, A., Lamers, M., & Bush, S. R. (2022). Contested mobilities in the maritory: Implications of boundary formation in a nomadic space. *Environment and Planning c: Politics and Space, 40*(1), 221–240. https://doi.org/10.1177/23996544211016866

Barrena, J., Bush, S. R., & Lamers, M. (2024). Countering salmon farming expansion: Network-making power in a nomadic marine space. *Geoforum, 148*, Article 103923. https://doi.org/10.1016/j.geoforum.2023.103923

Bollier, D., & Helfrich, S. (2014). *The wealth of the commons: A world beyond market and state*. Levellers Press.

Bottema, M. J., Bush, S. R., & Oosterveer, P. (2021). Territories of state-led aquaculture risk management: Thailand's Plang Yai program. *Environment and Planning c: Politics and Space, 39*(6), 1231–1251. https://doi.org/10.1177/2399654420978160

Brockington, D. (2004). Community conservation, inequality and injustice: Myths of power in protected area management. *Conservation & Society, 2*(2), 411–432.

Brondizio, E. S., Ostrom, E., & Young, O. R. (2009). Connectivity and the governance of multilevel social-ecological systems: The role of social capital. *Annual Review of Environment and Resources, 34*, 253–278. https://doi.org/10.1146/annurev.environ.020708.100707

Bulkeley, H. (2005). Reconfiguring environmental governance: Towards a politics of scales and networks. *Political Geography, 24*(8), 875–902. https://doi.org/10.1016/j.polgeo.2005.07.002

Büscher, B., & Dressler, W. (2007). Linking neoprotectionism and environmental governance: On the rapidly increasing tensions between actors in the environment-development nexus. *Conservation & Society, 5*(4), 586–611.

Bush, S. R., Oosterveer, P., Bailey, M., & Mol, A. P. J. (2015). Sustainability governance of chains and networks: A review and future outlook. *Journal of Cleaner Production, 107*, 8–19. https://doi.org/10.1016/j.jclepro.2014.10.019

Castells, M. (2000). *The rise of the network society* (2nd ed.). Blackwell.

Chaffin, B. C., Garmestani, A. S., Gunderson, L. H., Benson, M. H., Angeler, D. G., Arnold, C. A., Cosens, B., Craig, R. K., Ruhl, J. B., & Allen, C. R. (2016). Transformative environmental governance. *Annual Review of Environment and Resources, 41*, 399–423. https://doi.org/10.1146/annurev-environ-110615-085817

Cox, M., Arnold, G., & Villamayor-Tomás, S. (2010). A review of design principles for community-based natural resource management. *Ecology and Society, 15*(4), Article 38. https://www.ecologyandsociety.org/vol15/iss4/art38/

Cullingworth, J. B. (2002). *The political culture of planning: American land use planning in comparative perspective*. Routledge.

Cumming, G. S., Epstein, G., Anderies, J. M., Apetrei, C. I., Baggio, J., Bodin, Ö., Chawla, S., Clements, H., Cox, M., & Egli, L. (2020). Advancing understanding of natural resource governance: A post-Ostrom research agenda. *Current Opinion in Environmental Sustainability, 44*, 26–34. https://doi.org/10.1016/j.cosust.2020.02.005

Dell'Angelo, J., D'Odorico, P., Rulli, M. C., & Marchand, P. (2017). The tragedy of the grabbed commons: Coercion and dispossession in the global land rush. *World Development, 92*, 1–12. https://doi.org/10.1016/j.worlddev.2016.11.005

Djelantik, A. S. K., & Bush, S. R. (2020). Assembling tuna traceability in Indonesia. *Geoforum, 116*, 172–179. https://doi.org/10.1016/j.geoforum.2020.08.006

Doddema, M., Spaargaren, G., Wiryawan, B., & Bush, S. R. (2020). Fisher and trader responses to traceability interventions in Indonesia. *Society &*

Natural Resources, 33(10), 1232–1251. https://doi.org/10.1080/08941920.2020.1751129

Dupuis, E. M., & Vandergeest, P. (Eds.). (1996). *Creating the countryside: The politics of rural and environmental discourse*. Temple University Press.

Euler, J. (2018). Conceptualizing the commons: Moving beyond the goods-based definition by introducing the social practices of commoning as vital determinant. *Ecological Economics, 143*, 10–16. https://doi.org/10.1016/j.ecolecon.2017.06.020

Evans, D. (2012). Building the European Union's Natura 2000 network. *Nature Conservation, 1*, 11–26. https://doi.org/10.3897/natureconservation.1.1808

Folke, C., Hahn, T., Olsson, P., & Norberg, J. (2005). Adaptive governance of social-ecological systems. *Annual Review of Environment and Resources, 30*, 441–473. https://doi.org/10.1146/annurev.energy.30.050504.144511

Gómez-Baggethun, E., Kelemen, E., Martín-López, B., Palomo, I., & Montes, C. (2013). Scale misfit in ecosystem service governance as a source of environmental conflict. *Society & Natural Resources, 26*(10), 1202–1216.

Gotteland, E., Barrena, J., & Ibarra, S. (2024). Social transformations in marine conservation: The case of the Multiple-Use Coastal Marine Protected Area Pitipalena-Añihue. *Revista Austral De Ciencias Sociales, 47*, 7–26. https://doi.org/10.4206/rev.austral.cienc.soc.2024.n47-01

Hajer, M., & Zonneveld, W. (2000). Spatial planning in the network society: Rethinking the principles of planning in the Netherlands. *European Planning Studies, 8*(3), 337–355. https://doi.org/10.1080/713666562

Harambour, A., & Barrena Ruiz, J. (2019). Barbarie o justicia en la Patagonia occidental: Las violencias coloniales en el ocaso del pueblo kawésqar, finales del siglo XIX e inicios del siglo XX. *Historia Crítica, 71*, 25–48. https://doi.org/10.7440/histcrit71.2019.02

Harvey, D. (2012). *Rebel cities: From the right to the city to the urban revolution.* Verso.

Hein, L., van Koppen, K., de Groot, R. S., & van Ierland, E. C. (2006). Spatial scales, stakeholders and the valuation of ecosystem services. *Ecological Economics, 57*(2), 209–228. https://doi.org/10.1016/j.ecolecon.2005.04.005

Himes, A., Muraca, B., Anderson, C. B., Athayde, S., Beery, T., Cantú-Fernández, M., & Kenter, J. O. (2024). Why nature matters: A systematic review of intrinsic, instrumental, and relational values. *BioScience, 74*(1), 25–43. https://doi.org/10.1093/biosci/biad081

Holling, C. S., & Gunderson, L. H. (Eds.). (2002). *Panarchy: Understanding transformations in human and natural systems*. Island Press.

Hooghe, L., & Marks, G. (2003). Unraveling the central state, but how? Types of multi-level governance. *American Political Science Review, 97*(2), 233–243. https://doi.org/10.1017/S0003055403000649

Hull, V., & Liu, J. (2018). Telecoupling: A new frontier for global sustainability. *Ecology and Society, 23*(4), Article 41. https://doi.org/10.5751/ES-10494-230441

Kissinger, G., Brasser, A., & Gross, L. (2013). *Scoping study: Reducing risk: Landscape approaches to sustainable sourcing*. Landscapes for People, Food and Nature Initiative.

Mol, A. P. J., & Spaargaren, G. (2006). Towards a sociology of environmental flows: A new agenda for twenty-first-century environmental sociology. In G. Spaargaren, A. P. J. Mol, & F. H. Buttel (Eds.), *Governing environmental flows: Global challenges to social theory* (pp. 39–82). MIT Press.

Moore, M. (2012). Natural resources, territorial right, and global distributive justice. *Political Theory, 40*(1), 84–107. https://doi.org/10.1177/0090591711428188

Myers, N., Mittermeier, R. A., Mittermeier, C. G., da Fonseca, G. A. B., & Kent, J. (2000). Biodiversity hotspots for conservation priorities. *Nature, 403*(6772), 853–858. https://doi.org/10.1038/35002501

Ostrom, E. (1990). *Governing the commons: The evolution of institutions for collective action*. Cambridge University Press.

Ostrom, E. (2009). A general framework for analyzing sustainability of social-ecological systems. *Science, 325*(5939), 419–422. https://doi.org/10.1126/science.1172133

Pahl-Wostl, C., Gorris, P., Jager, N., Koch, L., Lebel, L., Stein, C., Venghaus, S., & Withanachchi, S. (2021). Scale-related governance challenges in the water–energy–food nexus: Toward a diagnostic approach. *Sustainability Science, 16*, 615–629. https://doi.org/10.1007/s11625-020-00888-9

Pellizzoni, L. (2018). Joining people with things: The commons and environmental sociology. In M. Boström & D. J. Davidson (Eds.), *Environment and society: Concepts and challenges* (pp. 281–304). Springer.

Pimbert, M. P., & Pretty, J. N. (1995). *Parks, people and professionals: Putting "participation" into protected area management*. United Nations Research Institute for Social Development.

Queffelec, B., Bonnin, M., Ferreira, B., Bertrand, S., Teles da Silva, S., Diouf, F., Trouillet, B., Cudennec, A., Brunel, A., Billant, O., & Toonen, H. (2021). Marine spatial planning and the risk of ocean grabbing in the tropical Atlantic. *ICES Journal of Marine Science, 78*(4), 1196–1208. https://doi.org/10.1093/icesjms/fsab006

Ribot, J. C., & Peluso, N. L. (2003). A theory of access. *Rural Sociology, 68*(2), 153–181. https://doi.org/10.1111/j.1549-0831.2003.tb00133.x

Sassen, S. (2006). *Territory, authority, rights: From medieval to global assemblages*. Princeton University Press.

Saunders, F. P. (2014). The promise of common pool resource theory and the reality of commons projects. *International Journal of the Commons, 8*(2), 636–656. https://doi.org/10.18352/ijc.477

Secretariat of the Convention on Biological Diversity. (2005). *Handbook of the Convention on Biological Diversity*. Earthscan.

United Nations Environment Programme. (2000). *Ecosystem approach: Description, principles and guidelines* (UNEP/CBD/COP/5/23; Decision V/6). UNEP.

UNESCO. (n.d.). *What are biosphere reserves?* Retrieved April 13, 2025, from https://www.unesco.org/en/mab/wnbr/about

Urry, J. (2000). *Sociology beyond societies: Mobilities for the twenty-first century.* Routledge.

van Koppen, K. C. S. A. (2006). Governing nature? On the global complexity of biodiversity conservation. In G. Spaargaren, A. P. J. Mol, & F. H. Buttel (Eds.), *Governing environmental flows: Global challenges to social theory* (pp. 187–219). MIT Press.

Vandergeest, P., Ponte, S., & Bush, S. R. (2015). Assembling sustainable territories: Space, subjects, objects, and expertise in seafood certification. *Environment and Planning A, 47*(9), 1907–1925. https://doi.org/10.1177/0308518X15599297

Vane-Wright, R. I. (1996). Identifying priorities for the conservation of biodiversity: Systematic biological criteria within a socio-political framework. In K. J. Gaston (Ed.), *Biodiversity: A biology of numbers and difference* (pp. 309–344). Blackwell.

Wang, J. H. Z. (2019). National parks in China: Parks for people or for the nation? *Land Use Policy, 81*, 825–833. https://doi.org/10.1016/j.landusepol.2018.11.054

Young, O. R. (2002). *The institutional dimensions of environmental change: Fit, interplay, and scale.* MIT Press.

Zuercher, R., Ban, N. C., Flannery, W., Guerry, A. D., Halpern, B. S., Magris, R. A., Mahajan, S. L., Motzer, N., Spalding, A. K., Stelzenmüller, V., & Kramer, J. G. (2022). Enabling conditions for effective marine spatial planning. *Marine Policy, 143*, Article 105141. https://doi.org/10.1016/j.marpol.2022.105141

9

Colonialism

Natalia Landívar

I write this chapter from territories covered under Treaty 8 and Treaty 6, negotiated between the Crown and Indigenous peoples in what is now known as Alberta, Canada. As an Ecuadorian scholar living and working in these territories, I write from a position that carries multiple, entangled layers of colonial histories—from the Spanish colonization of Abya Yala, known as Latin America, to the British and French settler colonization of Turtle Island, or North America, and their ongoing manifestations across both regions. In this dual positionality, I reflect on the parallels, continuities, and reconfigurations of coloniality in the Americas, exploring how these regions intersect in their historical experiences and contemporary understandings of colonial domination and resistance. Colonialism, in its broadest sense, refers to the political, economic, and cultural domination of one people over another through conquest, settlement, and control of territories and resources. It entails both material dispossession and epistemic subjugation, establishing hierarchies that privilege the colonizer's systems of knowledge, governance, and value over those of the colonized. The concept of colonialism clearly has substantive implications for socio-ecological relations generally, and for collective efforts to pursue just transitions specifically.

Historically, colonialism emerged as a global project of expansion initiated by European powers in the sixteenth century, organizing the world through what Peruvian sociologist Aníbal Quijano (2007) defined as the

N. Landívar (✉)
University of Calgary, Calgary, Canada
e-mail: Natalia.Landivar@ucalgary.ca

D. J. Davidson et al. (eds.), *Environment and Society*,
https://doi.org/10.1007/978-3-032-27059-7_9

"coloniality of power", or the racialized, hierarchical and Eurocentric organization of power, knowledge, and being that continues to structure the world. For world-system theorist Immanuel Wallerstein (2004), colonialism was not a process external to capitalism, but is a constitutive mechanism of global expansion, securing endless accumulation of capital through a core-periphery hierarchy. Similarly, Samir Amin (1988) conceptualized colonialism and imperialism as structural forms of unequal development that integrated peripheral regions into the capitalist system. While these analyses emphasized economic and geopolitical domination, postcolonial and decolonial scholars expanded the critique to include attention to the cultural and epistemic dimensions of colonial rule.

From the mid-twentieth century, postcolonial scholars examined colonialism as a system of cultural and psychological domination, not merely economic exploitation and political control. In The Wretched of the Earth, Frantz Fanon (1968) revealed how colonial power not only occupied land but also shaped the minds of the colonized, embedding inferiority through culture and language. Later, Edward Said (1978) exposed in Orientalism how the Western knowledge system produced and reproduced these hierarchies, representing non-Western societies as backward, irrational or dependent. Extending these insights, Wallerstein (2004) argued that modern knowledge is organized along colonial lines, with Europe (and its settler descendants) positioned as the locus of progress, rationality and universality.

Rather than a historical event confined to the past, colonialism has been reconfigured into enduring forms of domination that continue to shape global political, economic and ecological relations. In the Latin American context, Aníbal Quijano conceptualized this ongoing structure as coloniality, emphasizing the persistence of Eurocentric hierarchies of power, knowledge, and being that outlast formal colonial institutions. Building on the works of Quijano et al. (2018) argue that "there is no modernity without coloniality" (p. 4). Thus, modernity and coloniality are inextricably linked, as the narrative of progress and salvation is inherently tied to colonial violence. Nelson Maldonado-Torres (2007) further extends this framework through the notion of the "coloniality of being", which exposes how colonial logics dehumanize, racialize, and render entire groups as lesser forms of existence. Within Anglo-settler contexts, scholars such as Patrick Wolfe (2006) contend that settler colonialism also represents an enduring structure—one driven by a logic of elimination aimed at erasing Indigenous presence to secure permanent access to land and resources for settlers. Building on this insight, Lorenzo Veracini (2024) further conceptualizes settler colonialism as a distinct yet related formation of classical colonialism, maintained not through distant imperial control but through permanent occupation, Indigenous replacement, and the normalization of settler sovereignty.

Confronting coloniality through processes that have come to be known as decoloniality first emerged from Quijano's foundational work but was later theoretically elaborated and popularized by Mignolo in the early 2000s. The term serves as both an analytical lens and a praxis of delinking from the modern colonial order of knowledge, power, and being (Mignolo, 2007). It represents a project of re-existence grounded in pluriversality—the recognition of multiple ways of knowing, being, and relating to the world that have been suppressed or rendered invisible by Western modernity (Mignolo, 2007). Thus, delinking from Eurocentric hierarchies does not mean rejecting Western knowledge entirely, but rather challenging its universality, acknowledging that all knowledges are historically situated, politically mediated, and part of broader relations of power that have silenced other epistemic and ontological traditions (Fúnez-Flores et al., 2025).

Importantly, decolonial thought is not restricted to Latin America, but part of a broader global decolonial movement led by Indigenous, Afro-descendant, and other racialized peoples. As Maldonado-Torres (2025) notes, mobilizations surrounding the 500th anniversary of the so-called "discovery" of the Americas in 1992 marked a turning point in global anti-colonial thought and activism. These movements assert that decoloniality encompasses struggles and forms of resistance far beyond the Latin American context (Maldonado-Torres, 2025), and as Rivera-Cusicanqui (Fúnez-Flores et al., 2025) reminds us, they cannot remain a rhetorical or academic discourse, nor can they be reduced to state-appropriated slogans. In parallel, Black, Afro-diasporic, and Caribbean thinkers, such as Aimé Césaire (2000), Sylvia Wynter (2003), and Malcom Ferdinand (2022), have exposed how racial capitalism, environmental harm, and the transatlantic histories of dispossession and violence are constitutive of modernity itself, further pluralizing the genealogies of decolonial thought.

When we speak of decolonial thinking, or what in the Latin American context is known as decoloniality, it is essential to recognize its epistemological and geographical boundaries and the plurality of its genealogies, which emerge from diverse struggles against colonial violence and dispossession rather than from a single region or experience (Fúnez-Flores et al., 2025). Acknowledging this plurality allows us to connect decolonial thought with other traditions, including Black, Afro-diasporic, Caribbean, and Indigenous decolonial approaches present in the Americas.

In the North American context, Indigenous scholars approach decolonization from distinct epistemic locations that differ from the Latin American tradition of decoloniality. Their analyses emphasize resurgence as the moral, spiritual, and political foundation for decolonization (Alfred, 2005). For Tuck and Yan (2012), decolonization represents an unsettling and material project that involves the repatriation of Indigenous land and life, as well as the dismantling of settler-colonial structures. These scholars reject metaphorical or symbolic uses of the term, insisting that decolonization cannot be reduced to institutional reforms or academic discourses. Coulthard (2014) conceptualizes decolonization as a process of refusal and resurgence that rejects colonial recognition and reclaims Indigenous self-determination through land-based ethics and collective renewal. He grounds this process in "grounded normativity", the ethical and political practices of Indigenous peoples rooted in land, reciprocity and sovereignty. Similarly, Simpson (2017) defines decolonization from a perspective of the Nishnaabeg people as an individual and collective process of resurgence, rebuilding Indigenous knowledge systems and practices through reciprocal relationships with land, community, and interdependence. Together, these thinkers frame decolonization as a lived, relational, and ongoing process of regeneration of Indigenous freedom, rather than a metaphorical or state-led project.

Juxtaposing Latin American, Indigenous, Black, and Caribbean perspectives underscores that colonialism endures as a global structure, yet is contested through place-based, relational projects that exceed any single theoretical approach. By juxtaposing these perspectives, it becomes clear that decolonial thought is not a single, unified theory, but a constellation of epistemologies and place-based movements that confront the enduring global structure of colonialism from different historical and geographical locations. These diverse genealogies share a commitment to exposing how colonial domination continues to shape modern forms of life, knowledge, and economy, while simultaneously envisioning plural pathways towards justice and re-existence. In this sense, decoloniality and decolonization are not merely theoretical exercises but ongoing political, ethical, and ecological projects that redefine what transformation means in the contemporary world. Ultimately, how we understand and confront colonialism—across these geographies and epistemological traditions—will define the social, political, and ecological pathways of social transformation available to us.

It is from this plural, situated understanding of coloniality and decolonial thought that I turn to the contemporary energy transition. A rapidly expanding body of scholarship has begun to critically examine the contradictions and power asymmetries embedded in the pursuit of so-called "clean" or "green" energy. This literature reveals how colonialism is not only invoked as an abstract concept but also actively re-materialized in energy discourses, policies, and infrastructures, shaping which lands, knowledge, and futures are sacrificed in the name of transition (Mahony & Endfield, 2018; Whyte, 2017). Despite their promise of sustainability, many clean energy projects rely on the same colonial logics that have historically subordinated Indigenous, Afro-American and other racialized communities to the demands of global capital.

This chapter aims to identify the importance of attending to colonialism in the environmental social sciences, using energy transition politics as an illustrative case, and to propose a framework for decolonial social transformation. It develops an analytical lens composed of interrelated dimensions: epistemic hierarchies, top-down technocratic governance, territorial dispossession, ontologies domination, and resistance-alternatives. These dimensions reveal how the so-called clean energy transition remains embedded in the colonial matrix of power. By unpacking these dimensions, the chapter demonstrates how knowledge systems, governance structures, and material practices within energy transitions reproduce inequalities and forms of domination inherited from colonial rule. Building on this analysis, it argues that a truly sustainable transition requires more than technological innovations: it demands a profound transformation of social, epistemic, and ecological relations that underpin the modern/colonial order, moving towards decolonial futures grounded in reciprocity, care, and pluriversal understandings of energy and justice.

Coloniality in Energy Transition

The global energy transition is often celebrated as a pathway towards decoupling economic growth from fossil fuel dependence and associated greenhouse gas emissions and environmental degradation. Yet beneath its promise lies a profound contradiction: the transition itself depends on the intensified extraction of minerals, metals, and land to sustain its goal. According to the United Nations Environment Programme's Global Resources Outlook (2024), the extraction and use of natural resources have reached unprecedented levels in human history, increasing threefold over the past fifty years—from approximately 31 billion tonnes in 1970 to 95 billion tonnes in 2020. Far from reversing these trends, the shift towards "clean" energy is projected to accelerate further demand for so-called critical minerals and metals—including copper, lithium, cobalt, nickel, and rare earth elements—raising urgent concerns about new forms of environmental degradation, social conflicts, and deepening inequalities and injustices in Indigenous and resource-rich regions around the world.

Humans have always relied on nature in reciprocal and relational ways, even though dominant anthropocentric worldviews have long emphasized human superiority over, or separation from, the natural world. These dualistic perspectives—rooted in modernization theory— have obscured the interdependence between human and more-than-human life. One way to understand today's global climate breakdown and interconnected crises, as well as responses towards energy transition, is through the lens of extractivism.

Extractivism refers to a historical and ongoing mode of production and governance rooted in colonialism and reconfigured under neoliberal and neo-extractivist capitalism. It is organized around the large-scale removal of natural resources—such as minerals, oil, forests, land, and water—and the exploitation of human labour primarily for profit. It is not merely an economic process but a political, ecological, and ontological logic of domination that structures social relations, territories, and ways of life around the extraction of value from both nature and people (Rivera & Vindal, 2019). As Rivera and Vindal (2019) argue, extractivism operates through processes of enclosure and commodification that transform living landscapes into "resources", erasing Indigenous and local worlds in which mountains, rivers, and forests are understood as sentient and relational entities. This system concentrates wealth and power in the hands of corporations and elites while leaving local communities and ecosystems to bear the social and ecological costs. Furthermore, feminist scholars (e.g. Ojeda et al., 2024) describe extractivism as a gendered and colonial order that devalues life, enforces control over land and bodies, and reproduces hierarchies of power by prioritizing accumulation over care, sustainability, and justice.

Today, extractivism expands beyond material extraction to encompass new frontiers of accumulation, including carbon markets, biodiversity financialization, and the commodification of social cooperation and cultural identities, revealing extractivism as a pervasive mode of capitalist world-making. Green extractivism can be understood as the contemporary reconfiguration of extractivism, in which the large-scale removal of natural resources is reframed as a solution to the ecological crisis it helped to create. In this new form, resource extraction—particularly those deemed essential for "clean" or "renewable" technologies—is reframed as compatible with, or even indispensable to, climate change mitigation and environmental sustainability (Alkhalili et al., 2023; Dorn et al., 2022; Dunlap & Jakobsen, 2020; Riofrancos, 2023; Voskoboynik & Andreucci, 2022). Rather than representing a structural break from extractivist logic, these technologies—such as wind farms, solar parks, hydroelectric dams, lithium and rare earth mining, and carbon offset schemes—extend extractivist frontiers under a new moral and environmental guise.

Different authors have conceptualized green extractivism through distinct yet interrelated lenses. Voskoboynik and Andreucci (2022) describe green extractivism as a new discursive and material regime of accumulation, in which extraction is legitimized as environmentally sustainable and necessary to the global fight against climate change. Dunlap and Jakobsen (2020) situate green extractivism within what they call the "Worldeater" epoch—a totalizing techno-capitalist system that relentlessly expands the infrastructures of extraction, domination, and ecological control. Building on Svampa's (2019) notion of the commodity consensus, Dorn et al. (2022) identify a new climate consensus that embeds extractive expansion in the discourses of decarbonization and planetary salvation. Riofrancos (2023) broadens this critique beyond the Latin American context, showing how the onshoring of "clean energy" supply chains in the Global North reterritorializes extraction, transforming ecological transition into a national security project that reproduces extractivist dependencies under the guise of sustainability. These intertwined processes, often described as green colonialism, or climate coloniality (Bhambra & Newell, 2023; Lang et al., 2024), reveal how the pursuit of decarbonization can reproduce long-standing global hierarchies of power, labour and knowledge.

Against this backdrop, the remainder of the chapter employs an analytical framework that examines how coloniality is rearticulated through specific dimensions of the energy transition. Based on recent literature from the Americas, the following subsections will develop six interrelated dimensions that illuminate how colonialism continues to shape energy security discourses and practices.

Analytical Dimensions of coloniality in energy transition literature

Epistemic Hierarchies

This dimension refers to the unequal valuation of knowledge systems in the energy transition, in which Western, technocratic, and Global North epistemologies are privileged as objective and universal. At the same time, other forms of knowing, especially those of subaltern populations, are marginalized, discredited, and erased. These hierarchies very often obscure the underlying power relations and social injustices embedded in supposedly neutral climate and energy policies. By tracing the colonial genealogy of climate science, Mahony and Endfield (2018) demonstrate how imperial expansion and racial hierarchies historically shaped scientific authority and environmental knowledge. Through ideas such as climatic determinism, European science produced hierarchies of civilization, thus justifying imperial control and notions of European superiority. Similarly, Sultana(2022) extends this argument by demonstrating that climate coloniality persists today, materially through extractivism, carbon markets, land grabs, and "green" solutions that perpetuate dispossession and sacrifice zones, and epistemically, through Eurocentric, technocratic, and depoliticized forms of knowledge that silence subaltern epistemologies and local lived experiences.

Scholars have explored these epistemic hierarchies in different contexts. Velasco Herrejón and Friant (2022) demonstrate how wind-energy projects imposed in the Isthmus of Tehuantepec, Mexico, disregarded Indigenous epistemologies, reproducing and perpetuating colonial patterns of cultural domination. Similarly, in his analysis of the U.S. solar transition and the Black Lives Matter movement, Lennon (2017) demonstrates that the conception of energy as a quantifiable, mechanized, and apolitical object reproduces the same structures of epistemic and ontological domination inherited from colonial and racialized systems of power.

The call for a decolonial approach to energy transition involves epistemic justice that not only recognizes but also restructures the power dynamics that define whose knowledge systems count in energy transition. This entails integrating pluriversal technologies—energy systems co-produced through diverse ontologies—and relational worldviews that root technologies in autonomy and ecological reciprocity. For Lennon (2017), decolonizing energy requires redefining it as vital relationality—a living force embedded in social, racial, and environmental relations—rather than a depoliticized, commodified resource. Cedano-Villavicencio and Rincón-Rubio (2024) similarly argue that energy transitions are socially situated processes, shaped by local contexts, emotions, and relationships, rather than purely technical transformations. Collectively, these perspectives underscore the need to transform both epistemic foundations and material infrastructure to achieve a just transition.

Top-Down Technocratic Governance

An interlinked dimension to understanding the nexus of energy transition and coloniality is energy security governance, i.e. the structures, processes, and actors that determine how energy is produced, distributed, and secured. These arrangements are often shaped by colonial infrastructure and institutional legacies, which have historically defined who has the authority to set up policies, allocate financing and direct action. Such legacies embed hierarchical relations that privilege expertise and centralized decision-making, while marginalizing local communities, who are denied autonomy to define problems, solutions, and priorities.

Within this framework, the pursuit of energy security has been framed in technocratic and state-centric terms, narrowly tied to the fossil fuel era and premised on ensuring a stable energy supply for nation-states rather than the well-being of societies and ecosystems (Fouladvand et al., 2024; Kivimaa, 2024). Emerging from Cold War geopolitics, this conception linked energy interdependence with national security and economic growth, turning energy systems into instruments of control and stability. As Kivimaa (2024) notes, these historically entrenched frameworks continue to prioritize infrastructure protection and system reliability over social and environmental justice. In contrast, Fouladvand et al. (2024) demonstrate that such technocratic bias obscures the role of governance quality, participation, and institutional coordination in shaping energy security outcomes.

This legacy of centralized control continues to inform contemporary climate governance. Lang et al. (2024) refers to the "climate of coloniality" to describe how the Global North externalizes the social and environmental costs of its "green" development onto the Global South through extractivism, carbon offset schemes, and unequal trade relations. On her side, Tabassum (2021) shows how adaptation and mitigation mechanisms such as emission trading, the Clean Development Mechanism (CDM), or frameworks such as "Reducing Emissions from Deforestation and Forest Degradation in Developing Countries" (known as REDD/REDD +), promoted by the United Nations Framework Convention on Climate Change (UNFCCC) and international development institutions such as the World Bank or International Monetary Fund (IMF) have privatized environmental governance, commodified carbon and shifted the responsibilities for climate action onto developing countries. These mechanisms transform atmospheric carbon into a tradable asset, reinforcing what Bumpus and Liverman (2008) call "accumulation by decarbonization" whereby corporations and states in industrialized countries profit from climate mitigation projects, while expanding the reach of market-based governance. The result is a renewed form of carbon colonialism that preserves the energy and economic security of the North by displacing the ecological burdens of decarbonization onto the South.

Box 1. COP30 as an example of top-down climate governance?
COP30 in Belém marked the largest Indigenous participation in the history of the climate negotiations and produced several high-profile pledges. A central outcome was the Intergovernmental Land Tenure Commitment, which aims to recognize Indigenous rights over 160 million hectares of tropical forests by 2030. Brazil also advanced the recognition of quilombo territories, demarcated ten Indigenous lands, and secured renewed international funding for Indigenous, local, and Afro-descendant communities. New financial mechanisms, such as the Tropical Forest Forever Facility, pledged several billion dollars to reward forest protection, with a portion intended for Indigenous and local communities.

Despite these announcements, Indigenous delegates stressed that protection on paper does not guarantee safety on the ground. Violence against land defenders continued during the summit itself, with leaders from Brazil, Colombia, and Guatemala reporting escalating attacks. Many emphasized that global climate summits offer symbolic recognition but do very little to ensure actual protection for those defending forests and territories.

Climate finance outcomes were mixed. While new commitments on adaptation and just transition were introduced, most relied on voluntary indicators and lacked clear, accessible funding pathways. African pastoralist representatives warned that tripling adaptation finance by 2035 is too slow for communities already losing crops, livestock, and access to wild foods. Critics also noted that new forest finance mechanisms offer generous returns to outside investors while placing risks on communities.

Major gaps remained in the final decisions. Negotiators did not commit to a fossil fuel phaseout, did not include a roadmap to end deforestation, and removed references to the impacts of critical minerals extraction, even though many of the required minerals for renewable technologies lie on Indigenous lands. Indigenous leaders described these omissions as evidence of continued influence from powerful industry actors in global climate governance.

COP30 introduced new language affirming Indigenous rights, including free, prior, and informed consent, but leaders cautioned that only sustained political pressure will determine whether these promises lead to real protection. The summit ultimately revealed a familiar pattern in international climate politics: significant symbolic and territorial gains paired with structural barriers, uncertain implementation, and limited mechanisms to ensure the safety and rights of Indigenous peoples.

Source: Adapted from Sonam Lama Hyolmo, "What was achieved for Indigenous Peoples at COP30?" Mongobay (November 27, 2025).

Such arrangements illustrate how technocratic and neoliberal modes of governance translate environmental and energy challenges mainly into problems solvable through market instruments and expert-led interventions. Decisions about infrastructure, investment, and technology, despite their importance, are concentrated in international financial institutions, national bureaucracies, and the private sector, producing top-down forms of governance that obscure political, social and ethical questions of justice, autonomy and consent. As Hanaček et al. (2024) argue, a just transition requires that affected communities exercise self-determination and prior informed consent in energy and climate decisions, which are largely absent from the prevailing governance structure. By externalizing responsibility, these technocratic frameworks undermine the transformative potential of energy transition, reducing it to a depoliticized exercise. In contrast, emerging community-based initiatives reveal alternative pathways for democratizing energy governance. Rooted in citizen empowerment, local production, and co-management of renewable resources, these initiatives challenge the centralizing tendencies of technocratic governance. As Cuerdo-Vilches (2024) shows, energy communities can serve as models of collective resilience and self-determination, advancing decarbonization goals while enhancing social cohesion and reducing vulnerability. Such examples highlight that a just energy transition depends not on the consolidation of technical expertise or market efficiency, but on the redistribution of power and the recognition of communities as legitimate political actors in shaping their own energy futures.

Territorial dispossession

The notion of territorial dispossession captures how energy transition, while framed as solutions to the climate crisis, often reproduces colonial patterns of land appropriation, spatial exclusion, and violence. Rather than transforming extractive relations, renewable energy infrastructures such as wind, solar, hydro, and biofuels frequently extend the logic of accumulation by dispossession into new geographies. As Sheidel et al. (2023) explain, the global shift from fossil fuels to renewables, such as biofuels, hydropower, wind, and solar, has generated "renewables grabbing", or the large-scale appropriation of land, water, and resources in the name of decarbonization. Similarly, Lamhamedi and De Vries (2022) demonstrate that the rapid global expansion of renewable energy projects, particularly in the Global South, reveals a land-energy nexus—a structural tension between ecological modernization's win–win narrative and the realities of social inequalities and territorial displacement. These work collectives demonstrate that contemporary "green" development perpetuates settler-colonial power structures (Alkhalili et al., 2023), thereby normalizing territorial expropriation under the guise of climate action. DeBoom (2025) further deepens this analysis by introducing the concept of "climate necropolitics" to expose how entire places and populations become sacrifice zones, where the costs of decarbonization are displaced onto marginalized communities in the Global South, while the benefits accrue to European and elite actors.

In Latin America, scholars have revealed how green capitalism intensifies the territorialization of extractivism under the guise of a new environmental banner. Pinheiro and Nogueira (2025) observe that the energy transition has shifted from focusing on land appropriation to a broader and more complex process of territorial appropriation, as capital now seeks not only land but also subterranean, maritime, wind, solar, and water territories, which they call integral territorial grabbing. By analyzing the extraction of Nickel, Andreucci et al. (2023) develop the notion of decarbonization by dispossession to expose how low-carbon technologies lead to deforestation, pollution, and violent dispossession, transforming former colonies into "green sacrifice zones". Similarly, Andreucci, D., & Zografos, C. (2022) show that dominant climate governance operates through a dialectic of improvement and sacrifice, legitimizing violence against specific populations and territories deemed in need of improvement, while simultaneously rendering them expendable in the name of climate goals. These Latin American analyses foreground how green extractivism consolidates new regimes of territorial control that entrench the coloniality of power, deepen socio-environmental inequalities, and erase Indigenous and peasant ontologies of land.

Box 2. Wind-Energy and Territorial Dispossession in the Isthmus of Tehuantepec

In Mexico's Isthmus of Tehuantepec, one of Latin America's most important wind-energy corridors, Zapotec and Ikoots (Huave) communities have mobilized against large-scale wind farms that they argue are imposed on their territories without meaningful consultation. Although the region has become central to Mexico's renewable energy ambitions, local communities report that wind companies and authorities have pushed projects forward through opaque decision-making processes, exclusion of affected residents, and violations of the right to free, prior, and informed consent.

Community leaders explain that the rapid expansion of wind parks has altered communal land systems, undermined traditional fishing and agricultural livelihoods, and restricted access to coastal and lagoon areas used for generations. Many residents describe a pattern of territorial dispossession in which private companies obtain long-term land leases under unequal bargaining conditions, while communities face the social and ecological burdens of the industry. Reports also describe intimidation, criminalization, and threats against Indigenous land defenders who oppose the projects or demand fair negotiation processes.

Despite these pressures, communities across Juchitán, San Mateo del Mar, and Unión Hidalgo have organized assemblies, legal actions, and public protests to assert their territorial rights. Their resistance highlights a central dilemma in the energy transition: even renewable energy megaprojects can reproduce extractive dynamics when implemented through top-down governance structures that prioritize corporate timelines over Indigenous self-determination. The case of the Isthmus shows that decarbonization does not automatically align with justice unless land rights and community governance are central to planning.

Source: Adapted from Amnesty International, "Mexico: Community resistance to windfarms in the Isthmus of Tehuantepec exposes human rights risks of green transition" (2022).

In North America, similar dynamics of dispossession unfold through the language of net-zero and critical minerals. Scott (2025) demonstrates that Canada's pursuit of net-zero emissions is being used to legitimize a new wave of extractivism in Treaty 9 territory, where increasing demand for critical minerals, such as nickel and lithium, enables governments and corporations to justify large-scale extraction on Indigenous land, perpetuating settler-colonial logics of appropriation, depletion and subjugation. Reed et al. (2025) reveal that Canada's clean energy policies remain grounded in market-based "green" solutions that ignore the historical and ongoing dispossession of Indigenous nations, masking extractivism beneath the rhetoric of sustainable development. These cases highlight that territorial dispossession is not confined to the Global South. As in Latin America, Indigenous and local communities in Canada face the burden of decarbonization, while states and corporations continue to profit from the expansion of new extractive frontiers.

A decolonized understanding of territory requires reimagining the energy transition beyond capitalist and colonial logics of control. Rather than viewing land as an empty space for energy production, it must be recognized as a relational space of life, care and reciprocity. Lamhamedi and De Vries (2022) call for a "land-energy balance" that integrates tenure security, governance, and community participation, ensuring that renewable energy development advances spatial justice rather than displacement. Reed et al. (2025) argue that achieving truly clean and just transitions demands centring Indigenous jurisdiction and knowledge systems, treating decolonization not as a metaphor but as a practice of returning land and power. Such visions align with what Latin American initiatives (Pimbert & Borrini-Feyerabend, 2019) describe as territories of life, or spaces governed by relational ontologies that prioritize collective well-being, ecological stewardship, and sovereignty over capitalist accumulation.

Ontological domination

Energy models impose universalistic, technocratic worldviews or understanding of reality that devalue diverse ways of knowing and relating to the energy, nature, and territory, especially those rooted in Indigenous and racialized communities' cosmologies. Dominant energy transitions, framed as technical, apolitical, and universally "green" reproduce colonial and capitalist logics that erase plural ways of being, knowing, and relating to the world. These transitions do not simply replace fossil fuels with renewables; rather, they impose a singular Western ontology of energy, one that defines nature as a resource, time as linear, and progress as modernization. Such a mono-ontological worldview violently overwrites relational and territorial ontologies, those of Indigenous peoples, peasants, and other local communities who understand energy, land, and life as interdependent, spiritual, and communal. As Kellner (2025) observes, while the global shift to low-carbon technologies necessitates a massive expansion of mining, with 54% of new projects located on or near Indigenous land, existing governance frameworks reproduce colonial and Western approaches that ignore not only Indigenous rights and epistemologies but also their ontologies. Similarly, global energy research and development remain deeply shaped by Western anthropocentric ontologies that marginalize or erase Indigenous cosmologies and epistemologies of energy (Mazzone et al., 2023).

In North America, Schelly et al. (2021) argue that energy transition research and policy framed through Western concepts such as sociotechnical imaginaries fail to recognize that different communities, especially Indigenous Nations, operate from fundamentally distinct and sometimes incommensurable ontologies. While Western ontologies assume a single, human-centred reality, Indigenous ontologies, such as those of the Anishinaabe in the Great Lakes region, are grounded in relationality (reciprocal relationship among human and more-than-human beings) and sovereignty (the autonomy of all beings). This ontological difference is often misrepresented in research and policy. For example, the term vulnerability is widely used in Arctic adaptation research to describe Indigenous peoples' susceptibility to climate impacts, but this framing can be paternalistic, portraying communities as dependent on external interventions and reinforcing state authority over the Arctic (Cameron, 2012).

A growing number of scholars are collaborating with Indigenous leaders to document Indigenous ontologies of energy, illustrating how the disruption of energetic relations violates both ecological and spiritual laws. Manari Ushigua (in Mazzone et al., 2023) explains that for the Sápara Nation in the Ecuadorian Amazon, energy is a living knowledge, a dynamic, vital force flowing through all beings, including plants, rivers, winds, and minerals. The Sápara view the Amazon as a living organism where each element (fire, wind, water, soil) has its own energy, agency, and rules. Fire, often referred to as the flower of the earth, represents transformation and communication between human and more-than-human worlds. Similarly, Deniza Kawany Fulkaxo Cruz (also in Mazzone et al., 2023) describes how the Kariri Xoco and Yanawá People in Northeastern Brazil understand energy as Mecá Hiaidé, the vital force circulating through the four elements (air, fire, water and earth). All beings contain divine energy; human, animal, and mineral life are connected through spiritual and energetic interdependence. Rituals, such as songs, dances, and prayers, sustain and balance these energetic relations, reaffirming reciprocity with the natural world.

Moving beyond Western cultural imaginaries towards recognizing Indigenous ontologies transforms how we define and design energy systems—from efficiency and extraction towards reciprocity, kinship, and respect for the sovereignty of all beings (Schelly et al., 2021).

Kellner (2025) argues that a paradigm shift towards a relational approach is necessary. This is not merely a different ontological stance but a matter of justice. Combining Indigenous and Western ontologies within participatory frameworks is crucial to address extraction-related conflicts, yet its integration is only meaningful if embedded within a broader decolonization process that confronts historical land dispossession and power asymmetries. In this vein, Mazzone et al., (2023) propose the "cosmologies of energy" approach, a framework grounded in Indigenous oral narratives, art, and lived experience, that centres Indigenous ways of knowing and being, understanding energy not as a commodity or mechanical resource but as a vital, relational, and spiritual force connecting humans, non-humans, and the Earth. A just and plural energy transition requires learning from Indigenous cosmologies of energy, which reframe energy as relational and sacred, and reimagine sustainability as coexistence rather than control.

Resistance and Alternatives

Resistance movements emerge as central actors in the contestation and transformation of energy systems. Far from being barriers to progress, these struggles reveal the deep colonial and capitalist underpinnings of global energy transitions. As Temper et al. (2020) show, communities resisting extractive projects are not only defending their territories from dispossession; they are also shaping alternative visions of decarbonization that emphasize justice, participation, and relationality. Resistance thus becomes a constitutive force in the redefinition of energy futures—one that exposes how dominant transitions reproduce inequality while opening pathways towards decolonial and socially grounded transformations. In this sense, resistance is both a critique of technocratic, top-down approaches and a practice of collective self-determination that reclaims energy as a right and a relational practice rather than a market commodity.

In Canada, scholars have documented how Indigenous resistance to extractivism and industrial development is redefining governance, care, and adaptation in the context of the climate crisis. Gobby et al. (2022) demonstrate that Indigenous-led resistance constitutes an alternative form of governance that arises in response to the failures of formal state systems, creating spaces for new ecological and political relations. Willow (2016) conceptualizes Indigenous extrACTIVISM as both political and ontological, emphasizing that struggles against extractivism are simultaneously assertions of sovereignty and cultural survival. Similarly, Cameron et al. (2021) show that Indigenous communities across the Canadian prairies are enacting land-based responses that restore relations with the land as a living relative, integrating decolonization, knowledge resurgence, and community-led climate action. Together, these studies reveal that Indigenous governance and land-based resurgence are not reactions to crisis, but rather proactive, transformative practices that shape just and sustainable energy futures.

Across Latin America, communities are also combining resistance with visionary alternatives that challenge the coloniality of energy and development. In other words, they are moving beyond opposition to build local, participatory, and just energy transitions that embody relational ways of living and counter both fossil and "green" extractivism.

As Soler-Villamizar (2023) and Soler and Rankin (2022) illustrate, social movements in Colombia are advancing community-led initiatives based on solar, biogas, and agroecological systems, reclaiming energy as a common good. These practices are grounded in principles of autonomy, reciprocity, and care, connecting energy justice to food sovereignty and gender equality. Some authors (Del Bene et al., 2019) frame these experiences under the concept of energy sovereignty, which understands resistance as a decolonial and ecofeminist project that democratizes control over energy systems and knowledge. Similarly, peasant communities in Intag, Ecuador, are resisting extractivist dynamics by building communitarian energy projects, including small-scale hydropower and renewable initiatives rooted in solidarity economies, as well as reciprocal relations with nature (Lang et al. (2024)). According to Lang et al. (2024), these initiatives embody pluriversal and convivial technologies that reframe energy as part of the web of life, not as a commodity for growth. Similarly, Cedano-Villavicencio and Rincon (2024) show that community solar project in Temixco and Chiapas in Mexico thrives when they integrate social agency, emotion, and context-specific needs, rather than imposing top-down technological solutions.

Conclusion: Towards a Decolonial Social Transformation

This chapter has argued that the contemporary energy transition cannot be understood apart from the enduring structures of coloniality that organize power, knowledge, territory, and life across the Americas. Writing from the entangled geographies of Treaty 8 and 6 territories and Abya Yala, I have traced how colonialism persists not as a closed historical episode, but as a global formation reproduced through epistemic hierarchies, technocratic modes of governance, territorial dispossession, ontological domination, and the violence of green development. These dimensions show that many clean energy projects, policies, and narratives, however, urgently frame continue to rely on extractivist, racialized logics that render lands, bodies, and worlds expendable in the name of transition.

At the same time, this chapter has shown that coloniality is neither total nor uncontested. Across Indigenous nations in Canada, Latin American territories of life, Black and Afro-diasporic struggles, and community-based energy initiatives, we see concrete practices that refuse sacrificial territories and reconfigure energy as a common, relational and cared-for domain. These practices materialize principles central to decolonial thought: land back and Indigenous knowledges, collective governance, consent and self-determination, convivial and pluriversal technologies, and ontologies that recognize more-than-human beings as political subjects rather than resources. Thus, attention to the concept of colonialism is essential to current debates in the environmental social sciences regarding collective responses to the climate emergency. Doing so highlights the need for a decolonial approach to societal and energy transitions, ensuring that greening initiatives do not simply reproduce injustices associated with the current political and economic order.

Discussion Questions:

How does thinking about colonialism as an ongoing structure—rather than a past event—change the way we interpret contemporary environmental issues?

What everyday assumptions do we hold that might be shaped by colonial ways of organizing the world, even if we don't consciously recognize them?

What responsibilities arise for individuals and institutions once we acknowledge that contemporary environmental crises are rooted in ongoing colonial structures?

What assumptions about nature, energy, or human-nonhuman relations underpin mainstream energy systems? What would an energy system grounded in reciprocity, care, and respect for more-than-human life look like in practice?

What values or ways of living are expressed through collective resistance to extractive projects, and how might these values inform more just energy transitions?

How do these assumptions differ from relational or land-based worldviews held by many Indigenous or local communities?

How does treating land as a "site" for energy production differ from treating land as a living space that carries cultural, ecological, and relational meaning?

What happens when different ontologies—different ways of understanding and relating to the world—come into conflict in climate or energy policy?

How can community-based or land-based forms of energy governance challenge the idea that the only path forward is large-scale technological expansion?

Which knowledge systems tend to be treated as legitimate, scientific, or objective in climate and energy policy—and which are not? What are the consequences of this hierarchy?

References

Wasáse: Indigenous Pathways of Action and Freedom

Alfred, T. (2005). *Wasáse: Indigenous pathways of action and freedom.* Broadview Press.

Alkhalili, N., Dajani, M., & Mahmoud, Y. (2023). The enduring coloniality of ecological modernization: Wind energy development in occupied Western Sahara and the occupied Syrian Golan Heights. *Political Geography, 103*, Article 102871. https://doi.org/10.1016/j.polgeo.2023.102871

Amin, S. (1988). Accumulation on a world scale: Thirty years later. *Rethinking Marxism, 1*(2), 54–75. https://doi.org/10.1080/08935698808657801

Andreucci, D., García López, G., Radhuber, I. M., Conde, M., Voskoboynik, D. M., Farrugia, J. D., & Zografos, C. (2023). The coloniality of green extractivism: Unearthing decarbonisation by dispossession through the case of nickel. *Political Geography, 107*, Article 102997. https://doi.org/10.1016/j.polgeo.2023.102997

Andreucci, D., & Zografos, C. (2022). Between improvement and sacrifice: Othering and the (bio)political ecology of climate change. *Political Geography, 92*, Article 102512. https://doi.org/10.1016/j.polgeo.2021.102512

Barbosa, L. P., & Nóbrega, L. N. (2025). From the agrarian question to the territorial question: Green grabbing and the corridors of extractivist dispossession in Latin America. *Land, 14*(5), 1104. https://doi.org/10.3390/land14051104

Bhambra, G. K., & Newell, P. (2023). More than a metaphor: "Climate colonialism" in perspective. *Global Social Challenges Journal, 2*(2), 179–187. https://doi.org/10.1332/EIEM6688

Bumpus, A. G., & Liverman, D. M. (2008). Accumulation by decarbonization and the governance of carbon offsets. *Economic Geography, 84*(2), 127–155. https://doi.org/10.1111/j.1944-8287.2008.tb00401.x

Cameron, L., Mauro, I., & Settee, K. (2021). "A return to and of the land": Indigenous knowledge and climate change initiatives across the Canadian Prairies. *Journal of Ethnobiology, 41*(3), 368–388. https://doi.org/10.2993/0278-0771-41.3.368

Césaire, A., & Pinkham, J. (2000). *Discourse on colonialism.* NYU Press. http://www.jstor.org/stable/j.ctt9qfkrm

Coulthard, G. S. (2014). Introduction: Subjects of empire. In *Red skin, white masks: Rejecting the colonial politics of recognition.* University of Minnesota Press.

Cuerdo-Vilches, T. (2024). Energy communities as models of social innovation, governance and energy transition: Spanish experiences. In S. Sareen & A. Martin

(Eds.), *Geographies of solar energy transition: Conflicts, controversies and cognate aspects.* UCL Press.

DeBoom, M. J. (2025). Displaced decarbonization: Climate necropolitics and the contested spatialities of green hydrogen in Namibia. *Environment and Planning C: Politics and Space*, 1–20. https://doi.org/10.1177/23996544251332856

Del Bene, D., Soler, J. P., & Roa, T. (2019). Energy sovereignty. In A. Kothari, A. Salleh, A. Escobar, F. Demaria, & A. Acosta (Eds.), *Pluriverse: A post-development dictionary* (pp. 178–181). Tulika Books.

Dorn, F. M., Hafner, R., & Plank, C. (2022). Towards a climate change consensus: How mining and agriculture legitimize green extractivism in Argentina. *The Extractive Industries and Society, 11*, Article 101130. https://doi.org/10.1016/j.exis.2022.101130

Dunlap, A., & Jakobsen, J. (2020). *The violent technologies of extraction: Political ecology, critical agrarian studies and the capitalist worldeater*. Palgrave Macmillan.

Fanon, F. (1968). *The wretched of the earth*. Grove Press.

Fanon, F. (2001). *The wretched of the earth* (C. Farrington, Trans.). Penguin Classics.

Ferdinand, M. (2022). *Decolonial ecology: Thinking from the caribbean world* (A. P. Smith, Tran.). Polity Press.

Fouladvand, J., Sarı, Y., & Ghorbani, A. (2024). Infrastructure and governance: Prioritising energy security dimensions for community energy systems. *Energy Research & Social Science, 116*, Article 103676. https://doi.org/10.1016/j.erss.2024.103676

Fúnez-Flores, J. I., Beltrán, C. C. D., Ndlovu-Gatsheni, S. J., Bakshi, S., Lao-Montes, A., & Ríos, F. (2025). The SAGE handbook of decolonial theory. *SAGE Publications*. https://doi.org/10.4135/9781036204358

Gobby, J., Temper, L., Burke, M., & von Ellenrieder, N. (2022). Resistance as governance: Transformative strategies forged on the frontlines of extractivism in Canada. *The Extractive Industries and Society, 9*, Article 100919. https://doi.org/10.1016/j.exis.2021.100919

Hanaček, K., Kröger, M., & Martínez-Alier, J. (2024). Green and climate colonialities: Evidence from Arctic extractivisms. *Journal of Political Ecology, 31*(1), 538–566. https://doi.org/10.2458/jpe.5512

Kellner, E. (2025). Mineral extraction on Indigenous land: Employing a relational approach to navigate the convergence of Indigenous and other ontologies and practices. *Energy Research & Social Science, 125*, Article 104097. https://doi.org/10.1016/j.erss.2025.104097

Kivimaa, P. (2024). Energy security and geopolitics of energy transition. In *Security in sustainable energy transitions* (pp. 35–51). Cambridge University Press. https://doi.org/10.1017/9781009368155.005

Lamhamedi, B. E. H., & de Vries, W. T. (2022). An exploration of the land–(renewable) energy nexus. *Land, 11*(6), 767. https://doi.org/10.3390/land11060767

Lang, M., Bringel, B., Manahan, M. A., Svampa, M., Lander, E., Hamouchène, H., Feffer, J., Dietz, K., Escobar, A., Yanez, I., Moreno, C., Dorninger, C., Acosta,

A., Martínez, E., Brand, U., Bassey, N., Hertanti, R., Roa, T., Bertinat, P., … Fernandes, S. (2024). *Más allá del colonialismo verde: Justicia global y geopolítica de las transiciones ecosociales*. Universidad Andina Simón Bolívar.

Lennon, M. (2017). Decolonizing energy: Black lives matter and technoscientific expertise amid solar transitions. *Energy Research & Social Science, 30*, 18–27. https://doi.org/10.1016/j.erss.2017.06.002

Mahony, M., & Endfield, G. (2018). Climate and colonialism. *Wiley Interdisciplinary Reviews: Climate Change, 9*(2), Article e510. https://doi.org/10.1002/wcc.510

Maldonado-Torres, N. (2007). On the coloniality of being: Contributions to the development of a concept. *Cultural Studies, 21*(2–3), 240–270. https://doi.org/10.1080/09502380601162548

Maldonado-Torres, N. (2025). Aníbal Quijano and the decolonial turn. *Theory, Culture & Society*, 1–9. https://doi.org/10.1177/02632764241303701

Mazzone, A., Fulkaxò Cruz, D. K., Tumwebaze, S., Ushigua, M., Trotter, P. A., Carvajal, A. E., Schaeffer, R., & Khosla, R. (2023). Indigenous cosmologies of energy for a sustainable energy future. *Nature Energy, 8*(1), 19–29. https://doi.org/10.1038/s41560-022-01121-7

Mignolo, W. D. (2007). Delinking: The rhetoric of modernity, the logic of coloniality and the grammar of de-coloniality. *Cultural Studies, 21*(2–3), 449–514. https://doi.org/10.1080/09502380601162647

Mignolo, W. D., & Walsh, C. E. (2018). On decoloniality: Concepts, analytics, praxis. *Duke University Press*. https://doi.org/10.1515/9780822371779

Ojeda, D., Nirmal, P., Rocheleau, D., & Emel, J. (2024). Feminist ecologies. *Annual Review of Environment and Resources*.

Pimbert, M. P., & Borrini-Feyerabend, G. (2019). *Nourishing life—Territories of life and food sovereignty* (ICCA Consortium Policy Brief No. 6). ICCA Consortium.

Quijano, A. (2007). Coloniality and modernity/rationality. *Cultural Studies, 21*(2–3), 168–178. https://doi.org/10.1080/09502380601164353

Reed, G., de Almeida Cardoso, A. C., Alook, A., Young, A., Johnson, L. R., & McLay, K. (2025). Reclaiming energy transitions in Canada: Centering Indigenous peoples, rights, and knowledges for decolonial climate solutions. *Environment and Society, 16*, 130–154. https://doi.org/10.3167/ares.2025.160108

Riofrancos, T. (2023). The security–sustainability nexus: Lithium onshoring in the Global North. *Global Environmental Politics, 23*(1), 20–41. https://doi.org/10.1162/glep_a_00668

Rivera, J., & Vindal, C. (2019). Introduction: Indigenous peoples, extractivism, and turbulences in South America. In C. Vindal & J. Rivera (Eds.), *Indigenous life projects and extractivism: Ethnographies from South America* (pp. 1–50). Springer. https://doi.org/10.1007/978-3-319-93435-8_1

Said, E. W. (1978). *Orientalism*. Pantheon Books.

Scheidel, A., Sorman, A. H., Avila, S., Del Bene, D., & Ott, J. (2023). Renewables grabbing: Land and resource appropriation in the global energy transition. In A. Neef, C. Ngin, T. Moreda, & S. Mollet (Eds.), *Routledge handbook of global land and resource grabbing* (pp. 189–204). Routledge. https://doi.org/10.4324/9781003080916

Schelly, C., Gagnon, V., Arola, K., Fiss, A., Schaefer, M., & Halvorsen, K. E. (2021). Cultural imaginaries or incommensurable ontologies? Relationality and sovereignty as worldviews in socio-technological system transitions. *Energy Research & Social Science, 80*, Article 102242. https://doi.org/10.1016/j.erss.2021.102242

Scott, D. N. (2025). The power of 'net zero': Seductive dispossession on the critical minerals frontier. *Law Text Culture, 28*(1), 100–138. https://doi.org/10.14453/ltc.1717

Simpson, L. B. (2017). As we have always done: Indigenous freedom through radical resistance. *University of Minnesota Press*. https://doi.org/10.5749/j.ctt1pwt77c

Soler-Villamizar, J. P., & Rankin, A. J. (2022). Energías comunitarias para la transición justa. *Gestión y Ambiente, 24*(Supl2), 252–266. https://doi.org/10.15446/ga.v24nsupl2.92941

Sultana, F. (2022). The unbearable heaviness of climate coloniality. *Political Geography, 99*, Article 102638. https://doi.org/10.1016/j.polgeo.2022.102638

Svampa, M. (2019). Neo-extractivism in Latin America: Socio-environmental conflicts, the territorial turn, and new political narratives. *Cambridge University Press*. https://doi.org/10.1017/9781108752589

Tabassum, N. (2021). Climate capitalism or carbon colonialism? The critical features of climate change adaptation and mitigation policies. In *Handbook of development policy* (pp. 313–323). Edward Elgar Publishing.

Temper, L., Avila, S., Del Bene, D., Gobby, J., Kosoy, N., Le Billon, P., Martínez-Alier, J., Perkins, P., Roy, B., Scheidel, A., & Walter, M. (2020). Movements shaping climate futures: A systematic mapping of protests against fossil fuel and low-carbon energy projects. *Environmental Research Letters, 15*(12), Article 123004. https://doi.org/10.1088/1748-9326/abc197

Tuck, E., & Yang, K. W. (2012). Decolonization is not a metaphor. *Decolonization: Indigeneity, Education & Society, 1*(1), 1–40.

United Nations Environment Programme. (2024). *Global resources outlook 2024: Bend the trend—Pathways to a liveable planet as resource use spikes*. UNEP.

Velasco-Herrejón, P., Bauwens, T., & Calisto Friant, M. (2022). Challenging dominant sustainability worldviews on the energy transition: Lessons from Indigenous communities in Mexico and a plea for pluriversal technologies. *World Development, 150*, Article 105725. https://doi.org/10.1016/j.worlddev.2021.105725

Veracini, L. (2024). *Settler colonialism: A theoretical overview*. Palgrave Macmillan.

Villamizar, J. P. S. (2023). Community contributions to a just energy transition. In *Routledge handbook of Latin America and the environment* (pp. 449–459). Routledge. https://doi.org/10.4324/9780429344428-45

Voskoboynik, D. M., & Andreucci, D. (2022). Greening extractivism: Environmental discourses and resource governance in the "Lithium Triangle." *Environment and Planning E: Nature and Space, 5*(2), 787–809. https://doi.org/10.1177/25148486211006345

Wallerstein, I. (2004). Historical origins of world-systems analysis: From social science disciplines to historical social sciences. In *World-systems analysis: An introduction* (pp. xx–xx). Duke University Press.

Whyte, K. (2017). Indigenous climate change studies: Indigenizing futures, decolonizing the Anthropocene. *English Language Notes, 55*(1–2), 153–162. https://doi.org/10.1215/00138282-55.1-2.153

Willow, A. (2016). Indigenous ExtrACTIVISM in boreal Canada: Colonial legacies, contemporary struggles and sovereign futures. *Humanities, 5*(3), 55. https://doi.org/10.3390/h5030055

Wolfe, P. (2006). Settler colonialism and the elimination of the native. *Journal of Genocide Research, 8*(4), 387–409. https://doi.org/10.1080/14623520601056240

Wynter, S. (2003). Unsettling the coloniality of being/power/truth/freedom: Towards the human, after man, its overrepresentation—An argument. *CR: The New Centennial Review, 3*(3), 257–337. https://doi.org/10.1353/ncr.2004.0015

10

Knowledge

Rolf Lidskog and Göran Sundqvist

Introduction: Knowledge and Environmental Problems

Climate change, air pollution, chemical spills, and eutrophication. Carbon dioxide, PFAS, microplastics, and fine particles. We find these words in news articles, international political negotiations, and on the websites of environmental organizations. Words that signal that our planet is under threat from ongoing and escalating environmental pollution, that the situation is urgent, and that we must act swiftly to halt or at least mitigate environmental degradation. Otherwise, species will become extinct, polar bears will drown, places will become uninhabitable, and the health of the entire planet will be compromised as more planetary boundaries are crossed. The situation is so serious that we can no longer afford to address environmental problems slowly and incrementally, as we have in the past. Instead, we must rapidly develop and implement far-reaching societal changes. In short, society must transform. This transformation must be carried out fairly, ensuring that groups, communities, and countries are not affected unequally. Equity and

R. Lidskog (✉)
Örebro University, Örebro, Sweden
e-mail: rolf.lidskog@oru.se

G. Sundqvist
University of Gothenburg, Gothenburg, Sweden
e-mail: goran.sundqvist@sts.gu.se

D. J. Davidson et al. (eds.), *Environment and Society*,
https://doi.org/10.1007/978-3-032-27059-7_10

justice must be at the core of both knowledge production and its translation into action.

What characterizes all these environmental problems is their dependency on knowledge. Without understanding them, we cannot solve or manage them. We must first recognize an environmental issue and grasp its causes. Then, we need to develop knowledge about relevant, effective, and feasible solutions. Knowledge is necessary, but not sufficient, for action. History shows us many environmental problems that were well understood but took a long time to address (and in many cases, still lack strong responses).

Today, we possess extensive knowledge about climate change, biodiversity loss, and chemical pollution—the so-called "triple planetary crisis" (UNFCC, 2022)—yet action remains insufficient. Knowledge is growing, but the situation is deteriorating. The gap between knowledge and action is widening. This raises questions: what is knowledge, and what role should it play in transforming society toward sustainability? This chapter explores these questions.

We begin by outlining our sociological foundation: the sociology of knowledge. This perspective stresses that knowledge is not separate from society. It is primarily a practice, an activity, rather than a product, a fixed truth unaffected by social factors.

In the second section, we trace how classical sociology of knowledge has evolved into the sociology of science. We highlight that science is deeply embedded in society, and that modern life is unimaginable without science. Science has contributed to the creation of environmental problems, but it is also crucial for solving them.

The third section addresses the risk of placing blind faith in science as a problem solver. We examine the pitfalls of scientism—the belief that science is the only valid way to gain knowledge about the world marginalizes other forms of knowledge and dismisses non-scientific perspectives as irrational or biased. We also consider the opposite danger, populism, which downplays scientific knowledge and claims that "common sense" is sufficient. In this context, we explore scientific denialism, which paradoxically blends elements of both scientism and populism.

In the fourth section, we discuss how to navigate between scientism and populism using a sociology of knowledge approach. We present the ambition to democratize science and advocate for greater epistemological diversity, including perspectives from the Global South. We stress that knowledge, including expert knowledge, is about competence and position. This sociological understanding can enhance the role of knowledge in promoting a just transformation toward sustainability.

Finally, we conclude by briefly illustrating how climate change knowledge can be made more relevant to everyday activities and for a just transformation. For this to happen, knowledge must be better integrated into social practices.

Sociology of Knowledge

A common definition of knowledge is a set of justified beliefs about the world, where justification involves having good reasons for believing a claim is true (Pritchard, 2013). Although knowledge always has a subject—someone (a person or a group) or something (an institution, such as a university) possesses, develops and cultivates knowledge—it is justified by rational reasons. In this sense, knowledge is often viewed as separate and autonomous from society. It is a fixed and stable truth that individuals, groups, and even a whole society can embrace—or misunderstand or ignore. Classical sociology of knowledge problematizes this separated understanding of knowledge. A key example is found in the work of Émile Durkheim, who argued that all societies hold beliefs that are shared and taken for granted by their members. These shared beliefs, which Durkheim (1976) referred to as *collective representations*, form the foundation of any society because they bind people together. From the perspective of individual members, these representations are perceived as truths, or *social truths*, that is, shared ways of understanding reality. These truths emerge from and are sustained by social practices; they are collectively produced and maintained. The connection between social practices and knowledge is strong in every society. Rather than existing separately from society, these social truths are at the core of society.

Durkheim's ideas are foundational to the tradition that views norms and beliefs as socially constructed. According to this perspective, beliefs are acquired through interaction with significant others (Mead, 2015) and through primary and secondary socialization (Berger & Luckmann, 1991). As members of groups and societies, we both influence and are influenced by these interactions. Social belonging (family, friends, in-groups), involving relationships and interactions in various organizational contexts (such as workplace and voluntary associations) as well as through social media, plays a crucial role in shaping and sustaining our expectations, feelings, values and beliefs.

What individuals value, feel, and believe has a social origin (shaped by others) and a social orientation (directed toward others). We cannot live without this social belonging, and therefore social validation from the groups we seek to belong to becomes crucial. This makes it difficult for individuals to question what is taken for granted within their group, that is, shared knowledge (Klintman, 2019). The sociology of knowledge emphasizes that the socially constructed nature of beliefs also extends to knowledge. Not only are worldviews, beliefs, and attitudes fundamentally social, but so too are our understandings of what is considered true and what is recognized as knowledge. Social belonging provides what Berger and Luckmann (1991) refer to as a "plausibility structure" that makes the ideas and beliefs we hold appear reasonable and true.

The sociology of knowledge thus makes several central claims:

- Knowledge differs from beliefs in that it is socially justified and often taken for granted (naturalized) and therefore not typically subject to reflection.
- Knowledge justifies the social order by supporting our perception of the world as normal and natural.
- Knowledge is socially constructed and depends on social validation (institutional support), yet it is often presented as objective and independent of its social context.

These claims are, in many ways, uncontroversial. In everyday discussions. It is common to argue that a person's or group's knowledge is socially determined or at least socially constructed. When someone asserts something as true that we do not accept as true, we often explain this discrepancy by pointing out that in the social context in which the person or group lives, social factors have shaped their understanding and led them to regard it as knowledge. However, in everyday life, our approach to understanding and explaining knowledge is often asymmetrical. We may view others' knowledge as shaped by social conditions, while our own knowledge is to be justified by rational reasons.

However, we are sometimes capable of being symmetric and reflexive about our own position and the knowledge we hold and value. Living in a late modern society often requires rational justification (give reasons for) of one's position and truth claims, and efforts to support why the knowledge one holds should be considered valid (Giddens, 1994; for symmetrical and asymmetrical explanations, see Bloor, 1991).

Originally, classical sociology of knowledge focused solely on everyday knowledge and public knowledge—such as religious, moral, and commonsense understandings (social truths)—and had little to say about scientific knowledge (scientific truths). This changed in the 1970s, when the sociology of knowledge began to be applied to scientific knowledge as well.

Sociology of Science

The idea of autonomy dominates the understanding of scientific knowledge. Scientific activity is seen as rational and scientific results as true. Rational processes and true results are considered self-explanatory; they do not need sociological analysis since they are presumed to be uninfluenced by society. Only false knowledge is thought to require explanation.

The sociology of scientific knowledge (SSK), which further developed the ideas from classical sociology of knowledge, emerged as a response to the understanding of scientific knowledge as separate from society and thus left sociologically unexplained. Here it is argued that scientific knowledge, whether judged true or false, requires sociological explanation. This approach was presented as a symmetrical understanding of scientific knowledge: beliefs considered true or false are both to be explained by the same type of causes, including social causes (Bloor, 1991). This means that the process from scientific hypotheses to collectively sanctioned and accepted scientific truths becomes an object of sociological study. This view can be summarized by saying that knowledge is not inherently true but *becomes* true through social processes.

This should not be interpreted as a shallow sociological relativism. On the contrary, the fact that scientific knowledge is socially produced is what gives it its strength and is a prerequisite for it to become a generally accepted truth. Denying the social character of knowledge and viewing it as autonomous from society renders its lifeblood invisible (Bloor, 1991).

Originally, the SSK tradition (in the 1970s) was based on a sharp distinction between scientific knowledge and social factors, explaining the former through the latter. However, in the 1980s, a more dynamic view emerged in which knowledge and social order are seen as co-produced—they mutually influence each other (Jasanoff, 2004). This view dominates today's research, and the field has been broadened into what is now called Science and Technology Studies (STS).

As a significant development, Actor Network Theory (ANT) introduced a *network* concept that links knowledge and social factors: both are part of the same network. Scientific knowledge is no longer explained by social factors such as social interests; instead, the focus is on interaction and mutual relationships (Callon & Law, 1982). By focusing on networks, the idea of the autonomy of scientific knowledge becomes almost incomprehensible. If scientific knowledge is to be influential, it requires not fewer but more connections, which is about strengthening its position in the network and thereby being able to make a difference.

In practice, however, science is often rhetorically separated from society. This process of concealing network relations and processes of co-production is called *purification.* According to Bruno Latour (1993), the fact that scientific knowledge is often understood as autonomous stems from living in a "modern" world where separation and autonomy are seen as something desirable. In modern society, science is expected to be separated from politics, religion and popular beliefs. Modernity celebrates separation and clear boundaries, but in practice, in the actual production of scientific knowledge, this is neither possible nor desirable. In practical research we find the opposite of separation, that is, *hybridization* (Latour, 1993: 10–11).

These processes of purification and hybridization can be captured by the concept of *stage management* (Hilgartner, 2000). Frontstage refers to what is publicly presented, while backstage refers to what actually occurs in research practices: discussions about what works, how to handle uncertainties, and how to determine valid results. Frontstage presentations are often more purified than backstage activities. Studying these processes of purification and hybridization is crucial, not least the interaction between them. Latour argues that hybridization is the fundamental process, and purification is secondary. Hence, he concludes that "we have never been modern" (Latour, 1993).

The concept of *co-production* is closely related to ANT's network concept. It implies that scientific knowledge and social factors (such as interests) evolve through a joint process of mutual becoming. Co-production should be understood descriptively, assuming that science and politics are intertwined. Issues of separation, distance and autonomy must be understood from a basis of dependence and interaction (Hilgartner et al., 2015; Jasanoff, 2004; Latour, 1993). This aligns with ANT's distinction between primary and secondary processes. Scientific knowledge is embedded in a political context, which shapes what is prioritized and what receives attention. This means that ontology (what is) and normativity (what should be) are linked. Knowledge is the result of social activities but also serves as a building block in shaping society; for instance, determining which risks are considered more acceptable than others (Jasanoff, 2004).

Nevertheless, the idea of science as separate from politics remains strong and should be taken seriously. Much current research in STS focuses on the interplay between science and politics, challenging the influential *linear model of expertise*, which is "deeply embedded in both the science and the policy communities" (Sarewitz, 2000: 81). This model suggests that better expertise creates certainty, leads to better decisions and can resolve political disagreements. However, critical STS assessments conclude that more knowledge often leads to more conflicts and, consequently, inaction (Beck, 2011:

298, 303). The linear model assumes a separation and one-way influence between science and society; science influences society, but not vice versa (Soneryd & Sundqvist, 2023). In contrast, the concept of co-production stresses understanding actors within their cultural context. Knowledge is not an external force to be adapted to; it is shaped through social processes such as socialization and social belonging (Lidskog & Sundqvist, 2018).

Science and Politics

Although the linear model is questionable, it raises an important issue: how science influences—and should influence—politics and society. First it must be acknowledged that science and its applications (technology) are indispensable to social development. All societies depend on knowledge and are strongly shaped by it. For example, farming methods, plant and animal breeding, and road and house construction are all based on people's ability to generate knowledge from their practices and experiences. However, with the development of modern society, science has become a crucial infrastructure for everyday life. Although science has a long history, its growth, institutionalization, and professionalization are primarily a feature of modern society. Modern society is inconceivable without science and technology.

There is a scientific overflow in society, where knowledge and products generated in laboratories and research centers reach society with increasing speed and scope (Jasanoff, 2005). As a result, nearly all public issues have a scientific component. Additionally, the introduction of new technologies with uncertain implications has led to a growing number of scientific experts called upon by public administrations to assess risks and advise on courses of action. In this sense, we are witnessing a *scientization of society* in which nearly all aspects of life depend on expert systems (Giddens, 1991). The digitization of society exemplifies this reliance, where public services and commercial activities depend on digital technologies.

This has also led to a *scientization of politics*. Policymakers increasingly turn to expertise, either to guide action or to evaluate (or legitimize) decisions already taken. National governments and international policy bodies are establishing scientific organizations and expert panels with the explicit aim of providing a scientific basis for policy and assessing the impact of current policies (Lidskog, 2025). The most prominent of these are arguably the UN's Intergovernmental Panel on Climate Change (IPCC), which regularly assesses the scientific knowledge on climate change and develops advice for policymakers. This institutionalization of scientific advice requires the involvement of researchers and includes the belief that policies should be scientifically grounded, or at least scientifically credible.

These processes of scientization also imply a *politicization of science.* Scientific knowledge becomes part of political decision-making processes, and even when the scientific experts themselves believe that they are merely summarizing the state of knowledge and recommending ways to achieve established political goals, science becomes subject to public and political contestation (Weingart, 2023). This is evident when expertise is invoked in controversial issues, often with the aim of resolving these conflicts with the result of the conflicts intensifying.

Science as a Destroyer or a Savior?

Science and its applications are part of what is seen as both positive and negative in social development. They have enabled activities that are highly damaging to the environment, such as oil extraction, synthetic chemicals and environmental toxins. Yet, science has also played a vital role in solving environmental problems: catalytic converters that reduce certain car emissions, solar cells that convert sunlight into electricity, and alternative technologies that replace ozone-depleting substances.

This dual view of science—as both destroyer and savior—is central to the STS understanding of society's increasing reliance on expert knowledge and the parallel processes of scientization and politicization. This assessment is also reflected in environmental sociological theory (Lidskog & Sundqvist, 2018).

Ulrich Beck's theory of *the risk society* and *reflexive modernization* exemplifies the dual role of science (Beck, 1992, 2009). He describes how the development of the modern welfare society has unintentionally led to the creation of enormous risks that, if realized, could threaten the very existence of society. At the same time, Beck expresses strong confidence that science, if it adopts a reflexive approach, can help society to choose a different path, away from the nuclear, genetic and chemical mega-risks whose destructive potential threatens everyone, everywhere.

Another influential environmental sociological theory, *the treadmill of production* (Gould et al., 2008), distinguishes between *production science* and *impact science.* Production science is embedded in an institutional framework that promotes the expansion of goods and services (the treadmill of production), contributing to environmental degradation. Impact science, by contrast, develops knowledge about human impacts on ecosystems and seeks to halt environmental decline.

A third prominent theory, *ecological modernization*, emphasizes that science is already a driving force in creating sustainable social development (Mol, 2010). Through scientific research, new low-impact technologies, evolving regulatory frameworks, and growing environmental awareness, societies are beginning to modernize ecologically, replacing reactive, end-of-pipe solutions with preventive and proactive measures. Science thus plays a central role in environmental sociological theory. The approaches outlined above—through different emphases—all recognize the importance of science in addressing current environmental challenges and enabling societal transformations. Having contributed to environmental degradation and unsustainable development, science and technology are not only part of the problem but also part of the solution.

Between Scientism and Populism

Viewing environmental problems as fundamentally research-dependent—both in identifying them and in tracing their causes and proposing remedies—closely aligns with the notion of *scientism*. This term refers to the deeply entrenched belief that scientific knowledge is the only legitimate authority for justifying political action (Wynne, 2010). It places science at the center, functioning as an obligatory point of passage for environmental discussion and policymaking. Few environmental claims—whether made by governments, environmental movements, or private companies—are presented without reference to scientific knowledge.

Yet, science is often contested. The public may lack confidence not only in policies and regulations but also in the science underpinning them. Traditionally, the gap between scientific and public perceptions has been explained by the idea that citizens lack access to scientific knowledge and therefore draw incorrect conclusions. This explanation is known as *the deficit model of public understanding of science.* According to this model, the solution is to better inform and educate the public so that they possess the correct knowledge and can reach the same conclusions as the experts.

The deficit model has been heavily criticized, particularly by STS scholars (Irwin & Wynne, 2004). These scholars argue that the main problem is not public ignorance of scientific facts, but rather scientific experts' lack of awareness and interest in lay knowledge, how ordinary people assess situations when decisions must be made on complex and often controversial issues. Lay people should be welcomed into these discussions because they possess relevant knowledge, even in scientific and technical debates. This competence is not about scientific facts per se, but about the relevance and credibility of those facts, and about aspects that science may overlook. This kind of competence, which understands science within its institutional context, can enrich and provide a valuable contrast to specialized scientific and technical knowledge (Soneryd & Sundqvist, 2023).

This perspective on the importance of lay knowledge is fundamentally different from "science-related populism" (Mede & Schäfer, 2020) or "epistemological populism" (Collins et al., 2020), which claim that ordinary people, not scientists, possess the most relevant knowledge. This phenomenon is increasingly discussed, especially in relation to vaccination hesitancy (Kohler & Koinig, 2023) and climate denialism (Krange et al., 2018). Replacing blind trust in science with blind trust in lay people's beliefs, and disregarding scientific expertise altogether, draws a sharp boundary between experts and the public—asserting that one is right and the other is wrong. This growing polarization is at the heart of the post-truth society debate.

Scientific Denialism

The post-truth phenomenon, in which emotions and personal beliefs override knowledge, is often seen as a defining feature of contemporary society (Iyengar & Massey, 2019). However, knowledge resistance and scientific denial have a long history. In *Merchants of Doubt,* Naomi Oreskes and Erik Conway (2010) trace five decades of post-truth strategies and organized denialism, showing how small groups of scientists with close ties to industry and politics have misled the public. These actors promoted doubt and challenged studies linking smoking to lung cancer, coal burning to acid rain, CFCs to ozone depletion, and carbon dioxide emissions to global warming. Their strategy involved stressing scientific uncertainty, keeping scientific controversies alive, discrediting science and individual scientists, spreading misinformation, and demanding equal media coverage for their views.

Climate denialism is the most prominent example today. Studies show it is generated and supported by a "denial machine" of powerful vested interests that disseminate a message of widespread scientific disagreement and controversy about climate change (Dunlap & Brulle, 2020). This movement claims that it is being silenced, either by the mainstream scientific community protecting its own (biased) research or by a global conspiracy that favors the climate discourse (Dunlap & McCright, 2015). In contrast, organized denialism asserts that human-induced change is either not real or not problematic and portrays climate action as unnecessary, extremely costly, and harmful to society.

Social media significantly amplifies climate denial. Users may find themselves in echo chambers encountering only information that reinforces their views and forming like-minded groups with a particular framing of the climate issue (Cinelli et al., 2021). Feed algorithms further reinforce these dynamics, deepening polarization. In the US, the climate denial movement has created a parallel scientific universe, where counterarguments are widely disseminated through social networks (Dunlap & McCright, 2015). In such groups, the influence of established climate science, such as the IPCC reports, is undermined.

Climate denialism is critical in an asymmetrical way; it is skeptical to established science but uncritical to its own truth claims. Unlike the IPCC, whose assessments follow rigorous and transparent procedures for evaluating scientific knowledge (including how to classify and communicate scientific uncertainty), the denial movement merely claims to be based on "better" science without subjecting its own assumptions and arguments to critical scrutiny. Instead, it develops strategies to spread skepticism and foster blind trust among its followers (Boström et al., 2017).

Importantly, climate denialism is not driven by ignorance and lack of knowledge that can be corrected by better dissemination of knowledge. Instead, it is rooted in group identity and strategically orchestrated by powerful actors. The denial movement frames its message to appeal to people's fears and prejudices, sowing doubt about climate change and the need for action (Almiron et al., 2022; Dunlap & Brulle, 2020). Front stage facts, graphs, charts and stories of scientific misconduct are presented to convince people that climate change is a hoax. Backstage, however, investors, PR consultants, and think tanks are crafting strategies to promote skepticism and deny the seriousness of climate change.

Both scientism ("let science decide!") and populism ("let the people decide!") are deeply problematic. The former gives science a one-sided dominance, while the latter disregards scientific knowledge altogether. So, how does one navigate this landscape where scientism and populism compete for attention and trust?

Box 1. Climate, Climate Science and Climate Justice

Established as a United Nations body in 1988, the Intergovernmental Panel on Climate Change (IPCC) provides governments with scientific knowledge to inform climate policy. Through its comprehensive Assessment Reports (ARs), the IPCC synthesizes vast amounts of research to present a consensus view on climate change, its impacts, and potential responses. Produced by hundreds of scientists and reviewed by governments, these reports are widely regarded as authoritative. While the reports aim to be relevant to policy, they often avoid politically sensitive issues, which makes it easier for governments to agree on them.

Yet, despite the increasing sophistication and clarity of climate science, global greenhouse gas emissions have continued to rise. This paradox—more knowledge, yet insufficient political action—raises important sociological questions.

Expertise Between Science and Politics: Democratization and Political Influence

What does a sociological view of knowledge tell us about the relationship between science and politics? Can the symmetrical approach of sociology of knowledge help us to navigate contentious issues driven by a polarization between scientism and populism?

This tradition stresses that scientific knowledge is influenced by society. Science does not produce eternal, objective truths. Instead, it generates results that are often uncertain, provisional, and open to interpretation. Especially in the context of complex social problems, there are good reasons to question overly confident expert claims. History offers many examples of where scientific experts have raised expectations that were not fulfilled. Moreover, decision-making involves more than just expert knowledge. For many environmental issues, expert input is important, but not sufficient. Therefore, experts should not hold the power to make decisions on public matters.

To avoid technocracy based on scientism, a common proposal is to limit the power of science by increasing democratic and political influence, not only on public issues but also on scientific ones. Often, small groups of researchers, bureaucrats, and politicians collaborate in closed circles, presenting their work as scientific when it is, in fact, evaluative and political. These processes must be opened up to public scrutiny, exposing the political decisions that have been delegated to experts (Callon et al., 2009). The intended goal of inclusive strategies is to make better and more legitimate decisions by involving citizens and stakeholders in the production of knowledge (Nowotny et al., 2001). This kind of "democratization of science" is often welcomed in the public debate, and—at least rhetorically—also by many politicians and researchers (SAPEA, 2019).

However, there is a risk that such proposals may be misused to justify populist critique of decisions one disagrees with. STS scholars Harry Collins and Robert Evans (2007), while critical to absolute knowledge and scientism, argue that there must be limits to extended participation in knowledge processes and in decisions on technical issues. When scientism is criticized, there is often a call to broaden the scope of the debate and include more voices. But this can lead to a disregard for expertise and the quality of knowledge, resulting in flip-flop thinking where expertise is seen as either entirely good or entirely bad, swinging between scientism and populism.

In contrast, a sociological view of scientific knowledge encourages us to move beyond this all-or-nothing mindset. This perspective rejects the extremes of believing that scientific experts either can deliver absolute truths or have nothing valuable to contribute (Collins & Pinch, 1998: 155). Instead, it promotes a more nuanced understanding of science as a social process, shaped by context, values and interaction—yet still capable of producing valuable insights to inform democratic decision-making.

Box 2. Nuclear power, science and justice

Today, over 400 nuclear power reactors operate globally, and many governments consider nuclear energy essential for achieving fossil-free electricity production.

Proponents argue that nuclear power is a safe energy source with low greenhouse gas emissions, especially when compared to coal, gas and oil. They also emphasize technological solutions for managing nuclear waste (the spent nuclear fuel), asserting that it can be stored securely.

Critics, however, highlight the risks inherent in nuclear energy. Historical disasters—such as Harrisburg (1979), Chernobyl (1986), and Fukushima (2011)—serve as stark reminders of its potential dangers. Concerns also persist about the long-term safety of nuclear waste storage, with critics questioning the feasibility of guaranteeing secure containment over timespans reaching 100,000 years.

The debate is often polarized. Proponents may portray opponents as scientifically ignorant or resistant to technological progress, while opponents accuse proponents of being aligned with industry interests and overly reliant on technological optimism.

Knowledge Pluralism and Epistemologies of the South

The sociological discussion of knowledge, science, and the relationship between science and politics has primarily taken place within a Western context. Postcolonial theory has highlighted these limitations, stressing that the current environmental situation is deeply shaped not only by Western culture (how its members understand and orient themselves in the world), but also by Western science (the knowledge it produces about the world) (Santos, 2014). Postcolonial research stresses that many theories of social transformation are implicitly embedded in colonial ways of understanding society, where crucial power relations are obscured (Bhambra, 2014). These theories are often ethnocentric, evaluating the world from a particular position without recognizing or acknowledging that fact. The proposed remedy aligns with the democratization of science, advocating knowledge pluralism and epistemic diversity. This includes exposing the ethnocentric framing that conceives knowledge from the global South as a threat, ethnic minorities and indigenous communities as local or provincial, while positioning knowledge from the global North as universal. This calls for a symmetrical explanation of knowledge.

The inclusion of more diverse knowledge is not only a matter of representation or a corrective to the unequal racial and gender composition of science. It also involves broadening the knowledge base to include non-scientific forms of knowledge, such as traditional and indigenous knowledge. This pluralization raises epistemological challenges: how knowledge is managed, how it is judged to be valid, and how it becomes policy-relevant (White & Lidskog, 2023).

A prominent example of efforts to bridge different knowledge systems is the Intergovernmental Science-Policy Platform on Biodiversity and Ecosystem Services (IPBES) (Díaz et al., 2015). IPBES represents an innovative and ambitious institutional design for cross-fertilization and integration of different kinds of knowledge. However, the epistemological challenges of validating and synthesizing diverse forms of knowledge—especially non-scientific ones—remain unresolved (Löfmarck & Lidskog, 2017).

From the perspective of sociology of knowledge, this does not mean that non-scientific knowledge (such as indigenous knowledge) can be dismissed as particularistic and contextual in contrast to scientific knowledge. Rather, it stresses that even when science is staged and presented as universal, it shares with all knowledge the characteristics of being contextually influenced—it is co-produced. This does not imply that knowledge should not be critically examined and evaluated. Instead, it calls for a symmetrical understanding of how all knowledge is produced, maintained and questioned.

Such an approach requires subjecting one's own beliefs, ideas, norms and concepts to constant critical scrutiny. It involves recognizing the limits of our knowledge and practicing epistemic humility in the face of wicked or intractable problems (Lidskog et al., 2022). Moving society toward greater sustainability must be done reflexively, with attention to epistemic assumptions, normative commitments and the social embeddedness of the knowledge used to guide transformation. The search for *just* transformation makes it crucial to reflect on our often implicit and unarticulated normative commitments.

Expertise: Competence and Context

The fact that science and politics are not independent spheres in society does not mean that they collapse into one another or become indistinguishable. Instead, we should look for interdependencies, connections, and forms of integration, while remaining aware of strategies of separation and purification (Sundqvist & Linke, 2024). Similarly, the fact that knowledge production occurs in specific contexts that influence its content does not mean that knowledge can be reduced to its social context. We can—and should—continue to speak of knowledge and ignorance, of the epistemic and the social, of science and politics.

This becomes especially clear in the sociological concept of expertise. An expert is typically defined as someone with recognized and specialized competence in a particular field (Callon et al., 2009: 228). This competence—scientific knowledge, in the case of a scientific expert—is acquired through extensive training and socialization within a community of specialists (Collins & Evans, 2007).

No one becomes an expert in biology, medicine, or sociology in isolation; expertise requires mentorship, collaboration, and immersion of disciplinary and methodological practices. It is a social process, where competence is learned and certified by peers who judge and recognize one's competence.

Within these communities, some individuals are acknowledged as more competent than others, based on how well they perform their tasks. Thus, all experts rely on a broader community of specialists, both for their initial socialization and for ongoing recognition. Scientific experts must refer to what "the science says", and their authority is always collectively sanctioned (Barnes & Edge, 1982: 234). However, expertise also requires recognition and demand from the wider society. If specialist knowledge is acknowledged within a research community but not requested, trusted or valued by the wider society, its influence on broader decision-making will be limited. Becoming and being an expert always requires both internal legitimacy (within the expert community) and external trustworthiness (among those seeking advice). Thus, expertise is about both competence and social context, that is, it has both epistemic and social dimensions. The epistemic dimension refers to a content-related component, expertise is about competence (skills) that need to be acquired in learning processes. The social dimension highlights the contextual and relational aspects of how expertise is recognized and mobilized. It is also important to understand expertise as an institutional phenomenon. Although expert advice may be delivered by individuals, their competence and authority are shaped by—and are representing—institutions. These institutions must be trusted by society, and ultimately by citizens.

Climate denialism, as discussed earlier, illustrates how vested interests seek to undermine trust in scientific institutions and establish counter-expertise. This counter-expertise is framed as standing for "truth" in opposition to what is portrayed as corrupt science, bureaucracy, and politics. This means a strategic effort to erode public confidence in established institutions and to promote alternative narratives that serve specific ideological or economic agendas.

Conclusion: Embedding Science and Policy in the Wider Society

This chapter has stressed that knowledge, including scientific knowledge, is socially embedded, and that embeddedness is not a weakness but a strength. Knowledge derives its social power from the contexts of which it is produced and used. Yet this embedding is often obscured when scientific knowledge is presented publicly. While this may be intended to preserve scientific integrity and protect science from political influence, it is not an accurate portrayal—and it undermines the policy-relevance of scientific knowledge.

This can be illustrated by the case of climate research. The IPCC is widely regarded as the leading authority on climate change, and many expert organizations look to it as a role model (Hulme, 2024; Lidskog & Sundqvist, 2022). Through the IPCC, climate change is framed as a political issue requiring urgent multilateral agreements and coordinated action, with carbon emissions as the central focus. This framing has been successfully disseminated to political institutions, and few governments dispute it. In public discourse, scientific knowledge—often via media reports on the IPCC—is frequently invoked.

However, researchers disagree on the role of knowledge in policymaking. Some argue that climate change has been reduced to a scientific problem, centered on emissions, scenarios and technological solutions (e.g., Hulme, 2011). Others contend that the issue has been hijacked by fossil fuel interests and climate deniers (e.g., Mann, 2021). The former view suggests that science has become too dominant, while the latter says it lacks influence. This reveals flip-flop thinking with neither perspective addressing how climate research can be better integrated into policymaking and society.

As STS scholars stress, neither science nor policy should be analyzed in isolation. They are co-produced, part of the same cultural and institutional context, and should be understood as a package (Jasanoff, 2004). The IPCC produces assessment reports, but for this knowledge to become politically relevant, nation-states must translate it through their own cultures and incorporate it into their regulatory systems. Moreover, this knowledge must resonate with what citizens and other stakeholders find important, meaningful, and manageable. Without these connections, scientific proposals and political policies will lack public legitimacy and social acceptance. Scientific knowledge and expert advice do not exert influence independently of the cultures and networks in which they are embedded.

Expertise is not a silver bullet for solving social and political problems, but it is not irrelevant either. Climate change involves social, cultural, and political risks and choices, informed by what climate experts state regarding the environment and possible pathways forward. Yet, climate change also involves values and visions—how people see their society and what kind of future they strive for (Hulme, 2017; Wapner & Elver, 2016). In practice, proposed solutions are never value neutral. Each proposal excludes certain alternatives, reinforces existing societal trends, and favors specific technological approaches (Beck & Mahony, 2018; Berg & Lidskog, 2024). Some technical solutions, such as negative emission technologies (NETs), rely heavily on faith in unproven methods, such as carbon capture and storage (CCS). These proposals also tend to obscure unequal consequences—particularly for regions in the global South—such as shifting land use from food production to carbon sequestration through afforestation. This makes it crucial to engage with normative questions—for example, by asking: according to which criteria is a proposed pathway considered just? And for whom is it just? Keeping knowledge and context together is essential if knowledge is to play a role in policy development and in transforming society.

To conclude, this chapter has explored scientific and environmental knowledge as embedded in society by adopting a classical sociology of knowledge approach. This embeddedness makes knowledge a social force. Advancing the role of knowledge in guiding us toward sustainability does not require separating science from society. Rather, it requires deeper integration of knowledge into both individual and political decision-making processes. This involves incorporating environmental knowledge into the daily decisions and behaviors of individuals and organizations. It also means integrating scientific knowledge into society more effectively by building institutional support that makes this knowledge commonplace. Ultimately this will make the idea of a climate-neutral society seem self-evident and natural.

Discussion Questions:

Referring to the Box above, ***Nuclear power, science and justice,*** *consider how one might understand this conflict from a sociological perspective, for example by discussing this case in terms of knowledge claims, public trust, social values (such as equity), and power relations. To what extent do you find nuclear power compatible with a just transformation? If so, what is required to ensure that it meets the criteria of justice and security?*

Drawing on the sociological understanding of knowledge presented in this chapter, consider why knowledge tailored for political use has not resulted in adequate political action. What does this suggest about the role of expert knowledge in society and for policymaking? What strategies could help transform scientific truths about climate change into social truths—truths that are widely accepted, internalized, and acted upon by individuals, institutions, and governments? Would a stronger emphasis on politically sensitive issues, such as justice and equity, increase or decrease the relevance and the impact of the IPCC?

References

Almiron, N., Rodrigo-Alsina, M., & Moreno, J. A. (2022). Manufacturing ignorance: Think tanks, climate change and the animal-based diet. *Environmental Politics, 31*(4), 576–597. https://doi.org/10.1080/09644016.2021.1933842

Barnes, B., & Edge, D. (Eds.). (1982). *Science in context: Readings in the sociology of science*. The Open University Press.

Beck, S. (2011). Moving beyond the linear model of expertise? IPCC and the test of adaptation. *Regional Environmental Change, 11*(2), 297–306.

Beck, S., & Mahony, M. (2018). The politics of anticipation: The IPCC and the negative emissions technologies experience. *Global Sustainability, 1*, E8. https://doi.org/10.1017/sus.2018.7

Beck, U. (1992). *Risk society: Towards a new modernity*. SAGE.
Beck, U. (2009). *World risk society*. Polity Press.
Berg, M., & Lidskog, R. (2024). Global environmental assessments and transformative change: The role of epistemic infrastructures and the inclusion of social science. *Innovation: The European Journal of Social Science Research*. https://doi.org/10.1080/13511610.2024.2322642
Berger, P. L., & Luckmann, T. (1991). *The social construction of reality: A treatise in the sociology of knowledge*. Penguin Books. (Original work published 1967)
Bhambra, G. K. (2014). *Connected sociologies*. Bloomsbury Academic.
Bloor, D. (1991). *Knowledge and social imagery* (2nd ed.). University of Chicago Press. (Original work published 1976)
Boström, M., Lidskog, R., & Uggla, Y. (2017). A reflexive look at reflexivity in environmental sociology. *Environmental Sociology, 3*(1), 6–16. https://doi.org/10.1080/23251042.2016.1237336
Callon, M., Lascoumes, P., & Barthe, Y. (2009). *Acting in an uncertain world: An essay on technical democracy*. MIT Press.
Callon, M., & Law, J. (1982). On interests and their transformation: Enrolment and counter-enrolment. *Social Studies of Science, 12*(4), 615–625. https://doi.org/10.1177/030631282012004006
Cinelli, M., De Francisci Morales, G., Galeazzi, A., Quattrociocchi, W., & Starnini, M. (2021). The echo chamber effect on social media. *Proceedings of the National Academy of Sciences of the United States of America, 118*(9), Article e2023301118. https://doi.org/10.1073/pnas.2023301118
Collins, H., & Evans, R. (2007). *Rethinking expertise*. University of Chicago Press.
Collins, H., Evans, R., Durant, D., & Weinel, M. (2020). *Experts and the will of the people: Society, populism and science*. Palgrave Macmillan.
Collins, H., & Pinch, T. (1998). *The golem at large: What you should know about technology*. Cambridge University Press.
Díaz, S., Demissew, S., Carabias, J., Joly, C., Lonsdale, M., Ash, N., Larigauderie, A., et al. (2015). The IPBES conceptual framework: Connecting nature and people. *Current Opinion in Environmental Sustainability, 14*, 1–16. https://doi.org/10.1016/j.cosust.2014.11.002
Dunlap, R. E., & Brulle, R. J. (2020). Sources and amplifiers of climate change denial. In D. C. Holmes & L. M. Richardson (Eds.), *Research handbook on communicating climate change* (pp. 49–61). Edward Elgar Publishing.
Dunlap, R. E., & McCright, A. M. (2015). Challenging climate change: The denial countermovement. In R. E. Dunlap & R. J. Brulle (Eds.), *Climate change and society: Sociological perspectives* (pp. 300–332). Oxford University Press.
Durkheim, É. (1976). *The elementary forms of the religious life*. George Allen & Unwin. (Original work published 1912)
Giddens, A. (1991). *Modernity and self-identity: Self and society in the late modern age*. Stanford University Press.

Giddens, A. (1994). Living in a post-traditional society. In U. Beck, A. Giddens, & S. Lash (Eds.), *Reflexive modernization: Politics, tradition and aesthetics in the modern social order* (pp. 59–106). Polity Press.

Gould, K. A., Pellow, D. N., & Schnaiberg, A. (2008). *The treadmill of production: Injustice and unsustainability in a global economy*. Routledge.

Hilgartner, S. (2000). *Science on stage: Expert advice as public drama.* Stanford University Press.

Hilgartner, S., Miller, C. A., & Hagendijk, R. (Eds.). (2015). *Science and democracy: Making knowledge and making power in the biosciences and beyond*. Routledge.

Hulme, M. (2011). Reducing the future to climate: A story of climate determinism and reductionism. *Osiris, 26*(1), 245–266. https://doi.org/10.1086/661274

Hulme, M. (2017). *Weathered: Cultures of climate*. SAGE.

Hulme, M. (2024). "IPCC-envy": Why do other science-policy issues want an IPCC, and should they? *Dialogues on Climate Change, 11*(1), 7–11. https://doi.org/10.1177/29768659241293219

Iyengar, S., & Massey, D. S. (2019). Scientific communication in a post-truth society. *Proceedings of the National Academy of Sciences of the United States of America, 116*(16), 7656–7666. https://doi.org/10.1073/pnas.1805868115

Irwin, A., & Wynne, B. (Eds.). (2004). *Misunderstanding science? The public reconstruction of science and technology* (New ed.). Cambridge University Press. (Original work published 1992)

Jasanoff, S. (2004). The idiom of co-production. In S. Jasanoff (Ed.), *States of knowledge: The co-production of science and social order* (pp. 1–12). Routledge.

Jasanoff, S. (2005). *Designs on nature: Science and democracy in Europe and the United States*. Princeton University Press.

Klintman, M. (2019). *Knowledge resistance: How we avoid insight from others.* Manchester University Press.

Kohler, S., & Koinig, I. (2023). The effect of science-related populism on vaccination attitudes and decisions. *Journal of Behavioral Medicine, 46*, 229–238. https://doi.org/10.1007/s10865-022-00333-2

Krange, O., Kaltenborn, B. P., & Hultman, M. (2018). Cool dudes in Norway: Climate change denial among conservative Norwegian men. *Environmental Sociology, 5*(1), 1–11. https://doi.org/10.1080/23251042.2018.1488516

Latour, B. (1993). *We have never been modern.* Harvester Wheatsheaf.

Lidskog, R. (2025). Navigating global environmental challenges: Disciplinarity, transdisciplinarity, and the emergence of mega-expertise. *Climate, 13*(1), Article 20. https://doi.org/10.3390/cli13010020

Lidskog, R., Standring, A., & White, J. M. (2022). Environmental expertise for social transformation: Roles and responsibilities for social science. *Environmental Sociology, 8*(3), 255–266. https://doi.org/10.1080/23251042.2022.2048237

Lidskog, R., & Sundqvist, G. (2018). Environmental expertise as group belonging: Environmental sociology meets Science and Technology Studies. *Nature and Culture, 13*(3), 309–331. https://doi.org/10.3167/nc.2018.130301

Lidskog, R., & Sundqvist, G. (2022). Lost in transformation: The Paris Agreement, the IPCC, and the quest for national transformative change. *Frontiers in Climate, 4*, Article 906054. https://doi.org/10.3389/fclim.2022.906054

Löfmarck, E., & Lidskog, R. (2017). Bumping against the boundary: IPBES and the knowledge divide. *Environmental Science & Policy, 69*, 22–28. https://doi.org/10.1016/j.envsci.2016.12.008

Mann, M. E. (2021). *The new climate war: The fight to take back our planet*. Scribe.

Mead, G. H. (2015). *Mind, self, and society: The definitive edition.* University of Chicago Press. (Original work published 1934)

Mede, N. G., & Schäfer, M. S. (2020). Science-related populism: Conceptualizing populist demands toward science. *Public Understanding of Science, 29*(5), 473–491. https://doi.org/10.1177/0963662520924259

Mol, A. P. J. (2010). Ecological modernization as a social theory of environmental reform. In M. R. Redclift & G. Woodgate (Eds.), *The international handbook of environmental sociology* (2nd ed., pp. 63–76). Edward Elgar Publishing.

Nowotny, H., Scott, P., & Gibbons, M. (2001). *Re-thinking science: Knowledge and the public in an age of uncertainty*. Polity Press.

Oreskes, N., & Conway, E. M. (2010). *Merchants of doubt: How a handful of scientists obscured the truth on issues from tobacco smoke to global warming*. Bloomsbury Press.

Pritchard, D. (2013). What is this thing called knowledge? *Routledge*. https://doi.org/10.4324/9781315889443

SAPEA. (2019). *Making sense of science for policy under conditions of complexity and uncertainty* (Evidence Review Report No. 6). Science Advice for Policy by European Academies.

Santos, B. de S. (2014). *Epistemologies of the South: Justice against epistemicide.* Routledge.

Sarewitz, D. (2000). Science and environmental policy: An excess of objectivity. In R. Frodeman (Ed.), *Earth matters: The earth sciences, philosophy, and the claims of community* (pp. 79–98). Prentice Hall.

Soneryd, L., & Sundqvist, G. (2023). *Science and democracy: A Science and Technology Studies approach.* Bristol University Press.

Sundqvist, G., & Linke, S. (2024). Making science relevant: Comparing two science advisory organizations beyond the linear knowledge model. *Minerva, 62*(4), 527–547. https://doi.org/10.1007/s11024-024-09528-0

UNFCCC. (2022, April 13). *What is the triple planetary crisis?* https://unfccc.int/news/what-is-the-triple-planetary-crisis

Wapner, P., & Elver, H. (Eds.). (2016). *Reimagining climate change.* Routledge.

Weingart, P. (2023). Trust and distrust of scientific experts and the challenges of the democratization of science. In G. Eyal & T. Medvetz (Eds.), *The Oxford handbook of expertise and democratic politics* (pp. 29–51). Oxford University Press.

White, J., & Lidskog, R. (2023). Pluralism, paralysis, practice: Making environmental knowledge usable. *Ecosystems and People, 19*(1), Article 2160822. https://doi.org/10.1080/26395916.2022.2160822

Wynne, B. (2010). Strange weather, again: Climate science as political art. *Theory, Culture & Society, 27*(2–3), 289–305. https://doi.org/10.1177/0263276410361499

11

Conclusions

Debra J. Davidson, Magnus Boström, and Pedro Roberto Jacobi

Concepts are abstractions. They are useful, often powerful discursive orienting devices. As abstractions, however, they embody deeper assumptions and beliefs that too often reside beneath the surface. One particularly powerful use of concepts lies in their ability to enable better understanding of the dynamics of society-environment inter-relations, and even more so, concepts serve as guideposts in our collective efforts to address the more deleterious of those dynamics. These enablements, however, are not inevitable. The ready use of concepts whose problematic assumptions and beliefs remain hidden, unexamined, can serve to perpetuate the anthropogenic causes of environmental and climate crises, and send individual and collective action pursuits down fruitless and even problematic pathways. This volume seeks to invite the critical reflection needed to surface and examine those assumptions and beliefs, to ensure deliberate deployment of concepts with the greatest potential to elicit better understanding, and navigation of society-environment inter-relations, emphasizing theoretical approaches from

D. J. Davidson (✉)
University of Alberta, Edmonton, AB, Canada
e-mail: debra.davidson@ualberta.ca

M. Boström
Örebro University, Örebro, Sweden
e-mail: magnus.bostrom@oru.se

P. R. Jacobi
University of São Paulo, São Paulo, Brazil
e-mail: prjacobi@gmail.com

D. J. Davidson et al. (eds.), *Environment and Society*,
https://doi.org/10.1007/978-3-032-27059-7_11

a global perspective, with contributions from researchers of the Global North and South.

While the first edition of this volume sought to explore a set of concepts the critical scrutiny of which can enhance a better understanding of the dynamics of society-environment inter-relations in general, in this second edition, we narrow our gaze to focus squarely on a set of concepts that we perceive to be particularly relevant to shaping the collective navigation of those society-environment inter-relations, at a time when their just transformation is increasingly urgent across the world. In the years since publication of the first volume, we have seen extraordinary, disruptive, and in many cases unanticipated changes in everything from global economies, geopolitics and social movements, to our planetary systems. The invasions of Ukraine and Palestine, the rise of right-wing extremism across the globe but most emblematically embodied in the second election of U.S. President Donald Trump, have all posed direct challenges to previous progress in the pursuit of justice, multilateralism and social and ecological wellbeing. International climate agreements appear more precarious than ever. Meanwhile, the rapid escalation of the impacts of our climate and ecological emergencies are manifesting across the globe, just as projected by climate and environmental scientists, and in some cases even sooner than anticipated. In the academy, in part motivated by these emerging circumstances and also by the raising of new voices, our perspectives of society-environment inter-relations are once again in flux. Importantly, and constructively, emerging conceptual discussions are underscoring the intimate role of (in)justice in the productive of environmental and climatic disruptions, and confrontations with the outcomes of those disruptions. Emerging conceptual discussions are also upending human-centric notions of the distribution of agency among human and nonhuman actors, creating space for decolonial and non-western ecological worldviews.

The necessary pursuit of intentional, cooperative socio-ecological change could not be more clear. Just transformation has emerged as a powerful paradigm to govern such collective projects. As with so many other broad and encompassing social objectives, however, the efficacy and long-term outcomes of just transformation projects require zooming in, to draw attention to the assumptions and understandings—in other words, the concepts—that guide such projects. Concepts help us to evaluate what to sustain, and what to transform: by, for, and with whom.

This volume underscores the importance of attending to those concepts, critiquing their compatibility with just transformation objectives, and invoking them intentionally. Concepts also have the capacity to span social and institutional boundaries, building relationships that can foster cooperation and community. Concepts in support of just transformation must be able to accommodate complexity and transdisciplinarity, be embedded in society, attend to processes of change and resistance and to motivate innovative change-makers. Each of the concepts explored in this volume was selected based on their particular value to the project of just transformation.

The concepts chosen for reflection in this edition include those that attend to gross differences in power and wealth within and across territories, differences which can impose calamities upon both marginalized peoples and planetary systems alike: concepts like justice, and inequality. Close companion to these concepts is intersectionality, which highlights the multiple and complex ways in which inequality and injustice manifest according to social positionality. We also draw attention to colonialism, an underlying social structure that, until recently, has received far too little attention from the environmental social sciences, considering its intimate relationship with environmental and climate crises, and the exclusive character of our Western knowledge systems. Knowledge itself, also featured in our first edition, is once again brought in to our discussion of just transition here, noting in particular the rich bodies of traditional knowledge that have been actively silenced by colonialism.

Broader systems-level concepts like metabolism and commons, we argue, are equally germane to collective efforts to pursue just transformation, providing both an important filter through which to evaluate collective efforts to change material relations, as well as to reconsider those material relations not from the perspective of property, but as commons. We also seek to shift our gaze from linear and static depictions of economic systems to attend to the sociocultural intricacies of consumer demand, which scale up to collective "footprints." And finally, while our first edition devoted a chapter to risk, perhaps reflective of our own alarm at the rapid escalation in frequency and intensity of climate-attributed extreme events, we focus instead in this edition on disaster and strategies to cope with them. Living with disaster is now a defining feature of modern society. Disasters can become "normalized" to which communities respond with "business as usual" approaches, or they can serve as inflexion points: departures from normalcy that motivate creative rethinking of social agendas.

Through these concepts environmental social scientists can uncover the socio-environmental connections and processes involved in enriching certain people, communities and countries while imposing escalating hardship, pollution, destruction, and domination on human and nonhuman others—connections and processes that must be confronted to support an alternative future guided by just transformation. Just transformation is, however, not merely a vision and normative yardstick. Contributions of this book show that just transformation is a precondition for change: *a more equal world is necessary for sustainability transformation to happen.* This is so for several reasons. First, as shown in the inequality chapter, structural inequalities constitute drivers (directly and indirectly) of environmental decline and climate change. Second, the prioritization among civil societies of equality and justice is necessary for mobilizing public legitimacy and support for transformation. If change processes are driven without attention to interconnected inequalities (intersectionality) and racialized logics (colonialism), and without efforts to counteract current systems of domination locally and globally, these change processes are likely to do nothing other than further escalate conflicts and public criticism.

Third, publics may not feel encouraged or have the resources and capacities to participate in change-related practices and movements in situations where major inequalities and injustices permeate everyday life. Structural and interconnected inequalities feed feelings of hopelessness, powerlessness and cynicism. Fourth, sustained inequality and perceived injustices obstruct perceptions of shared responsibility and the establishment of trusting relationships and accordingly, inequality hinders broad collaboration and communication toward effective climate and environmental action. Such paralysis will continue at both the global and local scale, in the global South and global North.

We can also consider just transformation as a precondition for creating capacity for change; that is, cultivating broad public action-capacity for change-related work. The relation between various kinds of justices and the fostering of action capabilities is discussed in the chapter on justice. The dimensions of justice discussed in that chapter are enacted and negotiated through social movement actions. It is through broad, inclusive participation in change-related work that people and actors develop ideas and visions, and acquire the confidence, knowledge, and skills to engage. This social side of engagement facilitates the identification and experimentation of alternative practices within everyday life as well as in community initiatives of various kinds: protesting, sharing resources, developing community initiatives like community gardening, engaging in alternative economies. Thus, just transformation has the potential to facilitate a positive spiral of change. Action feeds action. Initial efforts lead to firm commitments. Specific change processes can be socially contagious: good ideas spread in and through social networks, communities, and movements locally and globally.

The creation of action-capacity is most powerful when transformation projects can be orchestrated into community initiatives and social movements, and become developed into 'commoning' projects, as emphasized in the chapter on the commons. Such collective action-capacity shows how agency is always embedded in distinct social contexts and physical places. Because climate change is interconnected with so many socio-economic activities, there is no single action pathway, but rather an entire ecosystem of possible entry points for action, constituting a plurality of pathways that can be selected to align with unique action capacities. The chapters in this volume demonstrate, for example, the importance of considering processes of learning, developments of cross-identity and cross-vulnerability activism (intersectionality chapter), and community-building efforts to support alternative sustainable practices. Chapters illustrate the importance of material and discursive spaces to support epistemic justice and self-determination, from which alternative frames, knowledge and worldviews can flourish; alternative frames with the potential to challenge dominant modes of thought that emanate from colonialist and technocratic/elitist mindsets, and facilitate a plurality of ways of thinking about our nature-society inter-relationships.

Crucially, the fostering of action-capacity for just transformation demands consideration of time and space. Even though our transformation need is urgent, change processes require persistence and long-term horizons. This appears as a temporal contradiction, but there is no paradox involved in the sense of urgency to take immediate action—based on existing knowledges and opportunity spaces—while using long-term visions and agendas. Contributions to this volume demonstrate the integral role of the spatio-temporal dimension shaping the experience of privilege, domination, identity, and opportunity—all crucial for creating incentives and capacities to take action.

These chapters also speak to the ways in which many of our social institutions are necessarily implicated in the work of just transformation. Despite its weaknesses, states, for example, remain an indispensable power container and regulator. The nation-state—as well as subnational and local public authorities—is key institutions through which many processes of transformative change must pass. Other institutions, including democratic structures, science, education, art, and media are likewise implicated. The chapters in this volume also speak to the transformative potential of knowledge systems and worldviews: feminist, anti-racial, decolonial, degrowth and commoning perspectives all challenge incumbent worldviews in ways that can create legitimacy for alternative ways of organizing and living. Chapters also address the importance of care ethics (for living and future generations, human and nonhuman life), as well as the primacy of principles of sufficiency and well-being over economic growth in policy and politics—all supporting alternative ideas and practices for bottom-up engagement.

Not the least, the paradigm of just transformation can help us identify and confront problematic structures and processes that prevent justice. This includes constantly drawing attention to how systems of domination are maintained, even embodied within purportedly green agendas. The chapters in this volume provide conceptual tools to identify, label, and scrutinize the many ways in which injustices are reproduced, their root causes, mechanisms, and manifestations, both in structures and discourses. Environmental sociology and related social sciences and humanities—through conceptual reflexivity—have a central role in revealing durable and deeply unsustainable socio-environmental interactions, connected to such deeply rooted processes such as capital accumulation, metabolic rift, unequal ecological exchange, consumerism, commodification of nature, technocratic governance, epistemic hierarchies, science denialism, imperialism and other structures of domination, destruction, extractivism, and violence. These structures and processes are all explored in this book, visualized and understood through the analytical toolbox of core concepts. Likewise, the same concepts can help reveal the many empty promises and hidden injustices contained in many ecomodernist agendas, such as green growth, green transition, green development, circular economies, and net zero policies.

Importantly, concepts give us the mental tools to engage in future-gazing. The very idea of a sociological imagination presumes both the capacity to see both how social relations have produced history, as well as how those social, and socio-ecological relations may unfold in the future. And yet, researchers across the social sciences and humanities have been reticent to engage in future-gazing—we are much more comfortable explaining the past. Our futures will inevitably be marked by potential, and potentially unavoidable, polycrises: scarcity of food and water, economic disruptions, and political conflicts to name a few. Scholars have a responsibility to grapple with those future scenarios. We need to offer insights under what conditions might we expect those polycrises to lead to conflict, and what conditions might facilitate cooperation, for example. This question is elemental, particularly with respect to relations between actors in the global North and global South, but there are important questions demanding future-gazing as well, such as how the distribution of risk may shift, and how the ruptures currently taking place across the global political economy can affect environmental governance capacities.

That future-gazing must also include reflections on our histories, however. The future has never been so uncertain, has never appeared to be so completely lacking in guideposts from history. Or is it? We have certainly endured catastrophic and disruptive epochs before, even some marked by climatic change. Such hindsight applied to our future-gazing can remind that collective action does indeed have the potential to support transformation. Through individual and collective actions, humankind has been able to establish democracies, human rights, welfare societies, institutions of health care and education, science, and many cultural institutions and practices that provide meaningful lives. The biggest lie for the future would be to claim that individual and collective action is impossible or futile. If people are taught that their actions are meaningless, they won't act—non-action and stalemate become self-fulfilling prophecies. Our task—the sociological imagination—is to free ourselves from the mindset that *what is* will remain so, to open the horizon of possibility for *what could be*. Concepts play crucial roles in opening this gaze for future-oriented collective action.

In other words, our attention must extend beyond the contours of our current polycrisis; we must also engage in future-gazing: envisioning pathways to move toward. Many social science and humanities researchers are equally loathe to venture here, although there are some notable exceptions, some of whom are featured in the chapters of this volume; scholars who offer conceptual frameworks to support decolonization, or degrowth, for example. But researchers are not the only ones seemingly reticent to venture here: environmental and climate movements themselves have all too often focused their discursive framing on projected depictions of futures to avoid, and relatively less effort crafting visions of futures to work toward.

We find ourselves at a crossroad, at which policymakers, researchers, business owners and communities must decide whether to continue to invest in a status quo that no longer has the capacity to support stability, indeed a status quo that for so many marginalized peoples never did provide stability, or to embark on a different pathway, one that will inevitably threaten incumbent powerholders, but offers a lifeline for the rest of us, and for the planet. Such a pathway demands investments in positive creativity and change-making. It is often through facing the precarious realities of life—including loss, job precarity, and the scarcity of resources—that the courage to create and act upon new pathways emerges, manifesting in collective efforts to support communities living with disaster, in new visions of inclusive governance arrangements, in the ontological courage to embrace multiple knowledges, and new worldviews. That positive creativity is far more likely to emerge when the hope and motivation to invest in such projects are prefaced by positive future visions. The concepts explored here can serve as feedstock for such future-gazing.

Index

D. J. Davidson et al. (eds.), *Environment and Society*,
https://doi.org/10.1007/978-3-032-27059-7

MIX
Papier | Fördert gute Waldnutzung
FSC® C083411

Kolibri 360 GmbH
Ferdinand-Jühlke-Straße 7
99095 Erfurt, Deutschland
produktsicherheit@kolibri360.de